Mad Dogs and Englishmen

Mad Dogs and Englishmen

Rabies in Britain, 1830–2000

Neil Pemberton

and

Michael Worboys

palgrave
macmillan

First published in 2007 by
PALGRAVE MACMILLAN
Houndmills, Basingstoke, Hampshire RG21 6XS and
175 Fifth Avenue, New York, N.Y. 10010
Companies and representatives throughout the world

PALGRAVE MACMILLAN is the global academic imprint of the Palgrave
Macmillan division of St. Martin's Press, LLC and of Palgrave Macmillan Ltd.
Macmillan® is a registered trademark in the United States, United Kingdom
and other countries. Palgrave is a registered trademark in the European
Union and other countries.

ISBN-13: 978–0–230–54240–2
ISBN-10: 0–230–54240–9

This book is printed on paper suitable for recycling and made from fully
managed and sustained forest sources. Logging, pulping and manufacturing
processes are expected to conform to the environmental regulations of
the country of origin.

A catalogue record for this book is available from the British Library.

A catalog record for this book is available from the Library of Congress.

Printed and bound in Great Britain by
CPI Antony Rowe, Chippenham and Eastbourne

Contents

List of Figures

List of Graphs and Map

Graphs

Map

List of Abbreviations

APRHAC	Association for Promoting Rational Humanity towards the Animal Creation
BMJ	British Medical Journal
BVA	British Veterinary Association
CD(A)A	Contagious Diseases (Animals) Act
DOPA	Dog Owners' Protection Association
EC	European Community
EU	European Union
LCC	London County Council
LGB	Local Government Board
MAFF	Ministry of Agriculture, Fisheries, and Food
MOH	Medical Officer of Health
NA	National Archives
NCDL	National Canine Defence League
PETS	Pets Travel Scheme
QRC	Quarantine Reform Campaign
RSPCA	Royal Society for the Protection of Animals
RVC	Royal Veterinary College
SPCA	Society for the Protection of Animals
SPH	Society for the Prevention of Hydrophobia and the Reform of the Dog Laws
WHO	World Health Organisation

Acknowledgements

The origin of this book was a chance conversation we had about neglected topics in medical history and the decision to explore the events behind the headline 'M. Pasteur v. The Chief Constable of Clitheroe', which appeared in the *British Medical Journal* in June 1890. Research, mainly in Lancashire newspapers, revealed an imagined contest over the treatment of rabid dog victims between the world's most famous scientist and the flamboyant head of police in a small East Lancashire town. On 28 January 1890 a rabid dog had run amok in towns east of Manchester; in Stalybridge it had bit four people and then another three in Hyde. The latter were sent to Paris to be treated at the internationally renowned Institute Pasteur, while the Stalybridge victims' treatment was organised by the Chief Constable of Clitheroe, who brought with him a herbalist from Colne – J. R. Hartley. The *British Medical Journal* presented a narrative of state-of-the-art science versus herbal hokum, made newsworthy by the death of a Stalybridge victim and the opportunity to attack unorthodox medical practice. In contextualising this story we soon realised that we had discovered a large, fascinating and, yes, truly neglected topic, not only in medical history but also in British veterinary, social, and political history.

In pursuing the many facets of rabies backwards and forwards in time, and in exploring other regions and the national scene, we have benefited greatly from the assistance of librarians, archivists, and fellow academics. We would like first to thank the librarians, archivists, family historians, and local historians in Manchester and Lancashire whose assistance revealed the richness of rabies as a research topic. In particular, we must acknowledge the help of Christine Bradley, Margaret Heap, Bill Jackson, John Strachan, and Craig Thornber. As the topic grew we benefited from the assistance of staff at the following libraries and archives: Special Collections at John Rylands University Library Manchester; Manchester Archives and Local Studies; Manchester Central Library; the British Library St Pancras; British Newspaper Library Colindale; National Archives Kew; the University of Sheffield Library; Sheffield Local Studies Library; the Record Office for Leicestershire, Leicester & Rutland; the West Yorkshire Archive Service; the Local Studies Library Bradford; Southampton Central Library; Exeter Central Library; Plymouth Central Library; Clitheroe Public Library; Stalybridge Public Library; Tameside

Record Office; Rochdale Public Library; Preston Public Library; the Central Library Halifax; the Royal Society for the Protection of Animals; the Kennel Club; the Dogs Trust; and Battersea Dogs Home. Next we must thank the following for permissions for illustrations: the Wellcome Library, London; the National Library of Medicine, Bethesda; the Cushing/Whitney Medical Historical Library, Yale University; PortCities Hartlepool; the RSPCA; Express Syndication; News International; and Lady Fretwell.

We have given talks on this topic to many groups, so we would like to thank all those who have listened, questioned, and argued with us in Baltimore, Bellagio, Cambridge, Hale, London, Manchester, Oxford, Warwick, and Washington. We have also benefited from specific advice from Abigail Woods, Pratik Chakrabarti, and Andrew Gardiner.

We would like to thank the Wellcome Trust for their support of this project and for their wider contribution to the history of medicine, which provides such a congenial community for work in the subject. The completion of the book was accelerated greatly by the time and facilities afforded to Professor Worboys by his Rockefeller Conference Center residency in Bellagio in October 2006. He would like to acknowledge the generosity of the Foundation, the excellence of the support from Pilar Palacio and her staff, and the generous advice from his fellow residents. Michael Strang and Ruth Ireland at Palgrave Macmillan expedited publication, and Vidhya Jayaprakash and S. Vidhya Shankar at Newgen Imaging in Chennai, India, provided efficient editing and production. Also, we must thank Jane Henderson for the index.

Last, but not least, we must thank our colleagues in the Centre for the History of Science, Technology and Medicine and Wellcome Unit for their tolerance of our enthusiasm for rabid dogs and for all their help and advice. In particular, we must acknowledge those who read and commented on the whole manuscript: Ian Burney, Vanessa Heggie, Rob Kirk, John Pickstone, Elizabeth Toon, and Duncan Wilson. Finally, we would also like to thank friends and family, especially Carole Worboys and Julia Worboys, for their critical reading of several chapters.

NEIL PEMBERTON AND MICHAEL WORBOYS

Introduction

Everyone can find dogs frightening. Almost all of us have been snapped at by a dog and many of us have crossed the road to avoid a potential confrontation with an aggressive animal. Now imagine that such dogs might be carrying a deadly disease, which, if you were bitten, might paralyse your body and unbalance your mind, before producing an inevitable agonising death. Rabies was and is such a disease. It was prevalent in Britain until its eradication in 1902, producing a regular death toll from its human form – hydrophobia. In this book, we return to the Victorian era when potentially rabid dogs lurked everywhere: at home, in the yard and on the street, in the press, in novels, in figures of speech, in popular memory, and in the imagination. The dread of rabies and hydrophobia was a constant presence and perpetual concern for the whole nineteenth century, and the threat of its re-emergence from imported animals continued throughout the twentieth century. The actual number of hydrophobia deaths was very small: only 1,225 were recorded between 1837 and 1902. But Victorians had to worry about any dog bite they received, and there were many because of the sheer number of stray and wild dogs around.[1]

Many commentators have observed that the public profile of rabies in Britain has been out of all proportion to its actual threat to health, but this misses the point that perceptions of risk are never rational and that, as this book will show, reactions to disease are socially and culturally revealing. In the late twentieth century, European states complained that the British government and its officials exaggerated the threat of imported rabies for political reasons, whilst the tabloids worried about foreign dog smugglers and foxes slipping into Kent through the Channel Tunnel. Indeed, Britain's rabies-free status became a feature of national identity; for example, it was used rhetorically by Margaret

1

Thatcher in the 1980s to exemplify the country's essential difference to Continental Europe – an island people, enjoying security and liberty behind secure borders. Yet, in popular culture the term 'Mad Dog' has become separated from rabies and associated with aggression and violence, as with the Ulster paramilitary leader Johnny 'Mad Dog' Adair, the 'bite-yer-legs' footballer Martin 'Mad Dog' Allen, and youth gangs like the Benchill Mad Dogs in south Manchester. Our medical and veterinary history of rabies is thus as much about the socio-cultural history of dogs and British identity, as it is about the understanding, prevention, and treatment of the disease.

This book is mostly about the Victorian era, with a final chapter on the twentieth century. We tell the story of how the incidence of rabies and hydrophobia waxed and waned over the nineteenth century, before their eradication in 1902.[2] We follow the interactions between medical, veterinary, government, and public knowledge and attitudes to rabies and hydrophobia, and explore the conflicts between these groups about how to control these diseases. Victorians were regularly reminded of the threat of rabies in popular memory, by word of mouth and through reports in newspapers. When rabies was present or feared, street posters warned of 'MAD DOGS' and 'HYDROPHOBIA', and instructed owners to keep their dog muzzled, on a lead or indoors. These notices were often posted during the Dog Days, the period from early July to early August that began with the rise of Sirius, the Dog Star, and coincided with the hottest time of the year. While rabies and hydrophobia were usually regarded as different forms of the same disease in different species, many professional and lay 'experts' held contrary views. Maybe they were quite distinct diseases, rabies in dogs being a physical disease and hydrophobia in humans being a mental disease. Or, perhaps, they were both imaginary, as many rabid dogs were assumed just to be aggressive, while many humans were thought to have the hysterical condition of spurious hydrophobia. By 1900, medical, veterinary, and lay opinion had closed around the view that rabies was a specific, contagious disease that was spread by the inoculation of a virus carried in saliva. Laboratory tests had made it possible to distinguish viral infection from spurious cases in fierce dogs and anxious people. Greater security had also been brought in 1885 by Louis Pasteur's preventive vaccine treatment for hydrophobia, which has an iconic status in medicine as being the world's first modern, medical breakthrough.[3]

In the twentieth century, most Britons saw rabies as a foreign, exotic disease which quarantine regulations kept out. It was associated first and foremost with dogs in Continental Europe and irresponsible owners who might reintroduce the disease by evading quarantines. However, in the 1920s rabies was increasingly experienced as a tropical disease by doctors working in the Indian or Colonial Medical Services and by Britons serving overseas. It was, of course, in this context that Noel Coward's 'Mad Dogs and Englishmen' song, first performed in 1932, found resonance. However, rabies was not 'tropical' because of climate; rather, it was present in locations where street dogs were tolerated and where people lived close to the wild animals that could transmit the disease. By the end of the 1930s, most industrialised countries had followed Britain and had dog rabies under control, but in the 1940s a new situation emerged as rabies went 'wild'. For example, in the United States the threat came from racoons and skunks and in Europe from foxes and wolves. In the early twenty-first century rabies continues to be prevalent in many countries and is estimated to cause around 50,000 deaths worldwide each year, 44 per cent in Africa, and 56 per cent in Asia.[4] If treated early enough, almost of these people could be saved, but rabies remains typical of many disease problems in poor countries, where the issue is the lack of resources and infrastructure to deliver services.

What *is* rabies? Well, we would prefer not to tell you at this point. We would rather you learnt what rabies *was* and how understandings changed with our historical actors, so that you reach the current state of knowledge at the end of the book. This way you will better appreciate past ideas and actions in context, and be less likely to interpret them through today's understanding. It is a fundamental requirement of historical scholarship that we think ourselves into the mindset of past generations and understand their world in their terms. Furthermore, it is essential that we do not regard past ideas and actions that are different to ours as simply wrong or foolish. This approach has to be adopted in the history of medicine in exactly the same manner as in other areas. So, with diseases we have to approach the views of past generations as we do approach their views of politics or religion, to be understood *relative* to time and place, and to be explained in context, not judged against modern knowledge.[5] For example, understanding tuberculosis in the nineteenth century requires that we discuss it as the inherited affliction that Victorians knew it to be, not as the communicable disease we know today. It can be harder to be historically relativist about medicine because of the assumption that knowledge of the body and disease is cumulative and progressive; in other words, today's knowledge is 'right' and closer to the 'truth' than that of the Victorians. But remember that today's scientific experts on rabies are also 'relativists'. They accept that their knowledge of the disease is changing and will change;

indeed, most are research scientists working towards that very end! That said, if you cannot wait and want to know the current state of play with rabies in animals and humans, then this is discussed in Chapter 6. In addition, and in the spirit of historical relativism, we have included online sources for following future changes, should you be reading this book some years or decades hence.[6]

Why rabies?

Why have we written a book on rabies? Indeed, why a book just on rabies in Britain – a country where the disease was stamped out a century ago, and where before then it was comparatively rare?[7] Our answer is that this narrative reveals important yet neglected features of the history of disease and medicine, and of British social and cultural history. Specifically, we highlight four themes: the relationships between human and veterinary medicine, the interactions of professional and popular understandings of disease, the role of state in controlling disease, and the changing place of the dog and dog ownership in British society.

Diseases that are communicable from animals to humans, and bridge veterinary and human medicine, have recently attracted a lot of attention, for example, Salmonella, variant CJD, and Avian influenza.[8] Many such conditions are emerging diseases, either truly novel or newly recognised, but infections transmitted from animals to humans also seem to be new because until recently they were not a serious danger to human health. They tended to be few in number, not transmit easily, and to be confined to certain groups. For example, anthrax was only caught by workers in the wool industry, glanders by those who spent a lot of time with horses, and psittacosis by those who sold or kept parrots.[9] Also, few of us worry about catching diseases from our pet cats and dogs. The great livestock diseases – rinderpest, foot-and-mouth, and pleuropneumonia – do not affect humans; indeed, the main threat was and is food poisoning from meat.

In the Victorian era the boundary between animal and human health was not so secure; indeed, the pioneering work after 1870s on how germs caused infection began with studies of diseases that crossed species barriers – anthrax, tuberculosis, and, of course, rabies.[10] But none of these conditions was straightforwardly 'catching': anthrax and rabies were spread by the inoculation of poison-germs through the skin, while tuberculosis and rabies had long, variable incubation periods. Also, such diseases often produced distinct symptoms in animals and man; for example, in rabies dogs were thirsty while in humans they were hydrophobic. Thus, through the history of rabies we can analyse how doctors and veterinarians understood and dealt with complex problems that were at the margins of their professional practice, and to the struggles over who had reliable knowledge and hence authority.[11] Rabies also reveals the problems posed to public policy by diseases with variable patterns of spread and development, and where it is not always clear who has appropriate expertise. Dog fanciers and the public often presented themselves to be as 'expert' as vets, doctors, and scientists, even after Louis Pasteur showed rabies to be a germ disease and introduced an anti-rabies vaccine. This is somewhat paradoxical as Pasteur's work has been celebrated by medical scientists as revolutionary, producing the first fruits from experimental laboratory investigations of disease, the type of medical research that has dominated medical research since.[12]

Our second theme is the differences between professional and lay understandings of disease. Throughout the nineteenth century doctors claimed that the gap between their knowledge and popular understandings was nowhere greater than over hydrophobia. For example, doctors admitted that it was the one disease for which they could offer no useful treatment once symptoms developed, whereas the public tried all manner of therapies, from literally applying the 'hair of the dog' to taking the remedies offered by local chemists. The idea that rabies was associated with heat, thirst, and perhaps solar influences persisted amongst many social groups. One problem was that there was no medical consensus on hydrophobia until the 1890s; indeed, there were many different groups producing understandings from different starting assumptions, by different means and to different ends. For example, doctors observed, treated, and wrote about a fatal disease in individuals, which some saw as wholly physical, others psychological, and others both. Thus, rabies has much to tell us about changing ideas about the relations between body and mind, especially phobias and what we now term psychosomatic illnesses. Veterinarians saw rabies in individual dogs, and as an epizootic – an imported animal plague – in the canine population.[13] But rabies had other 'experts': the police who had to control dogs on the street, social reformers who saw the disease as a metaphor for the culture of the poor, animal welfare activists who were certain it was caused by cruelty, dog fanciers and owners who had their pet theories about breeds and gender, government officials who saw rabid dogs as a proxy for actual or potential social disorder, and, of course, the public who knew, it seems, a 'mad dog' when they saw one, and almost certainly knew someone who had an infallible remedy. Knowledge was often geographically or socially specific, and meanings given to the disease and its treatments were contingent. We will use the notion that there were many 'experts' to show the social basis for the struggles over the nature of rabies and its management.

Our third theme is the role of the state in the control of animal and human diseases. The contests over the appropriate measures for rabies were part of wider public debates over the extent of government intervention in the private lives of its citizens, the values of a liberal society, and the politics of class that were taking shape in the reconfiguration of the meanings, forms, and boundaries of the nation and the polity. In the first attempts to introduce dog controls in the 1830s, legislation was rejected in part because the public saw the muzzled dog as symbolic of political oppression at the critical moment of the Reform crisis of 1830–32.[14] This was also a time when the English people celebrated the ferocious bull-dog as their icon.

How to control rabies was also shaped by gender politics. Rabies was typically male and associated with Englishmen rather than women – with street life, cruelty to dogs, and aggression. But at the end of the century, women led the fight against the compulsory muzzling of dogs, pointing to the innocence and passiveness of lap dogs and seeing the muzzle to be as oppressive as the corset and a symbol of male domination. They were offended that sporting dogs – in the male worlds of hunting, coursing, and dog racing – were exempt from muzzling. The taming of rabies was in part the story of the desired and idealised character and qualities of the English, as well as about individual freedoms and liberties. It reminds us how in the Georgian era Englishness was famed for its aggressiveness and tenacity – symbolised in the figure of John Bull and the blood sports enjoyed by men of all classes. Over the Victorian era this was replaced by a mild-mannered and tamed temperament, associated with the middle classes, who were proud of their sensitivity towards animal suffering and over time led the British to become a nation of dog lovers. These shifts have been discussed for the late Victorian period by John Walton in a pioneering article, which remains the best introduction to the social history of dogs and rabies.[15] We aim to extend his analyses both back and forward in time, and to deepen it using the tools of the social history of disease and medicine.[16]

Our fourth theme is the history of dogs in Britain. We cannot answer the question why and how Britain became a nation of dog lovers, but we do offer insights into changing attitudes towards dogs and the changing character of dog ownership. Throughout the nineteenth century battles raged over the sight and treatment of animals, over the proper place, use and treatment of draught animals and livestock, and over the alleged cruelty to animals in popular culture, particularly sport and entertainment.[17] There are no histories of the dog and pet-keeping in Britain, though a number of studies of animals consider the subject. Keith Thomas's magisterial *Man and the Natural World* charts the development of domestic pets as part of his larger study of the waning of anthropocentrism and the growth of sentimental attitudes towards animals, for example, giving dogs names and keeping them in the home as companions. Thomas attributes this development, first seen amongst the wealthy, to the declining economic role of animals and the separation of domestic life from immediate contact with the exploitation and killing of animals. In addition, he points to the influence of radical Christian sects, natural philosophy, and the Enlightenment in general, all of which combined to open the emotional and social space for the sentimentalisation of animals. However, the extent to which any section of society was insulated from contact with working animals and livestock, even after 1800 when Thomas's study ends, is a moot point. All kinds of beast were omnipresent in Victorian cities, and links between town and country remained close. In the case of attitudes to dogs after 1830, we have found every shade of opinion from those, exemplified by owners of lapdogs, who regarded their pet as equal to or above humanity, not least through the virtues of loyalty and affection, to those, typified by owners of draught dogs, who treated their animals as mere economic assets.

Harriet Ritvo's discussion of dogs in *The Animal Estate* remains the most detailed and convincing account of the place of dogs in Victorian culture.[18] Sources for the history of dogs are not extensive, though Ritvo skilfully uses 'uncommon' phenomena on which documentation is rich, such as dog shows and rabies, to illuminate the 'common' and everyday. Her main point, like Thomas, is that attitudes to animals were never simply that, but were also about people and society, and hence, were shaped by, and in turn shaped, cultural ideas and actions. In the case of Victorian Britain, Ritvo highlights the link between dogs and social class, with dog shows a symbol of social divisions and distinctions, and rabies associated with the lower orders, 'unsettling social forces', and programmes of social discipline.[19] Rabies was also a resource for metaphorical reflection on the 'self'; Victorians, like Charles Darwin, wrote about having 'rabid' feelings, not least on matters of scientific controversy, and novelists such as George Eliot and Anthony Trollope had characters who behaved like mad dogs.[20] Indeed, more has been said about the rhetorical uses of rabies than on the disease itself, or attempts at control.[21] We endeavour to follow Jonathan Burt's appeal to historians of animals and human–animal relations to 'move away from emphasis on the textual, metaphorical animal ... to achieve a more integrated view of the effects of the presence of the animals'.[22] Our approach to dogs, dog–human relations, and the management of dogs in the Victorian period stresses the materiality of dogs and their diseases, and the situatedness of knowledge and practices in time, place, and social relations.

Mad dogs and Englishmen

The narrative of the book moves in broad chronological order through the prevailing understandings of rabies and the measures taken to control the disease, beginning in 1830 with the 'Era of Canine Madness' and ending in 2000 with the introduction of Pet Passports. We focus mainly on England. In fact, most rabies outbreaks and hydrophobia deaths were in London, Lancashire, and Yorkshire, though Wales, Scotland, and Ireland will figure as necessary and at certain points were critical – the last known human death in Britain from indigenous rabies was in South Wales in 1899. We identify and discuss many constructions and meanings of rabies, which can be characterised by period. In Chapter 1, Rabies Raging, we discuss the growing problem of rabies and hydrophobia in Britain in the 1820s and its significance in the crisis year of 1830. In Chapter 2, Rabies at Bay, we cover the period from 1831 to 1863 when the incidence of the diseases declined, but show how they remained important in the popular imagination, being kept there by remembered experiences, local authority dog control campaigns, and above all by the actions of animal welfare reformers. In Chapter 3, Rabies Resurgent, we show how the return of rabies brought tougher controls and new understandings of the disease in dogs and humans. In Chapter 4, Rabies Cured, we follow the development and introduction in Britain of Louis Pasteur's preventive treatment, from initial scepticism to acceptance a decade later. In Chapter 5, Rabies Banished, we tell the story of the control of rabies in the 1880s and 1890s, which through ever stricter measures led to its eradication in 1902. In Chapter 6, Rabies Excluded, we discuss how rabies was kept out of Britain in the twentieth century by the application of rigid quarantines on imported dogs, cats, and other mammals. Unbending measures remained in place until 2000, when controls were relaxed with the introduction of Pet Passports for dogs whose owners have had them tagged with a microchip, vaccinated, tested, and certificated.

1
Rabies Raging: The 'Era of Canine Madness', 1830

In June 1830, the Home Secretary, Robert Peel, received a letter with the following observation:

> In the eventful history of this Kingdom there have occurred few calamities of a more appalling nature than those recorded by Hydrophobia, particularly within the last few weeks – indeed, the future historian may not aptly distinguish it by the 'Era of Canine Madness' that began to show itself some years previously – and annually increased until 1830 – when it was hardly safe to walk abroad.[1]

The writer was reflecting on a situation where London and other towns were in the grip of a 'great and almost universal alarm' about mad dogs and hydrophobia.[2] The press was full of harrowing reports of fearful scenes, bloody injuries, frightful symptoms, and tragic deaths. In London 'thousands and tens of thousands of dogs kept by the Poor' were reported to be roaming the streets, snapping and biting at anyone and anything in their path – people, other dogs or the other animals that inhabited the crowded thoroughfares. There were similar reports from around the country.[3] Everywhere stray dogs were rounded up or shot on the street. (See Figure 1.1.) T. L. Busby's cartoon published in 1826 shows two armed groups converging on a mad dog and conveys the horror and humour of such episodes.

All reports claimed that the nation's dog population was out of control and that public order was in jeopardy. Canine madness had belonged to the working-class streets and the dogs that infested them, though it was now spreading to respectable streets and squares, threatening homes, shops, and schools – according to the *Evening Mail*, the 'grim monster'

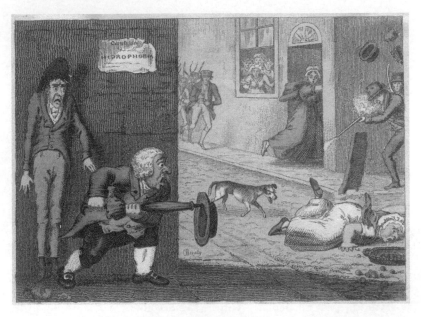

Figure 1.1   T. L. Busby, 'Mad Dog', 1826.[4] Courtesy of the Cushing/Whitney Medical Historical Library, Yale University.

had the country 'in its horrific sway'.[5] Thus, 'canine madness' referred not only to a disease in individual dogs, but also to the feeling that the canine population of London, major towns, and perhaps the whole country was suffering from a collective mania and out of control.

Reports for 31 May 1830 illustrate events and the public mood. At Bow Street Court a superintendent reported that 'the number of dogs in the streets without owners was frightful' and that a four-year old had been bitten on the lip. At Queen's Square Magistrates' Court there were reports of several people bitten in York Street the previous day, and no sooner had the magistrates' proceedings opened than they were interrupted by a dog bite victim. He had been attacked earlier that morning, had been to Westminster Hospital to have his wound 'cut out', and had come to court to ask if anything could be done to require owners to restrain or muzzle their dogs. Within an hour there was another application from a York Street victim, who had been bitten along with six others. The magistrates were considering a confinement order when a beadle entered the court with a dog; however, it was not the alleged rabid dog but with an ownerless cur, of which he said there were hundreds in the district. Confinement orders were introduced which allowed stray dogs to be rounded up, owners reclaiming dogs to be fined £5, and unclaimed dogs to be destroyed. A writer to *The Times* on 4 June asked 'who is there among us ... that leave his home in the morning, and say that he may not return in a few hours, brought back in a state that would reduce him to the desperation, and phrenzy (*sic*) of a demon, and from which a horrible death can alone relieve him?'[6]

The prospect of a terrifying illness and death was bad enough, yet any victim had to endure the surgical removal of tissues around the bite, cauterisation of the wound, and weeks of worry over whether they would be the unlucky one in ten, or was it one-in-hundred, that developed hydrophobia. A man from Hackney wrote to Peel pleading for government action after his son had been bitten by a dog that ran into the school playground, 'my fine boy had been tortured by cutting and cauterising, but at the moment he may be hanging between life and death, & oh! such a death'.[7] The next day Peel received a letter that urged him to give attention to 'the state of misery in which thousands are kept at their (perhaps unfounded) apprehension of mad dogs. This, with many accounts of agonising terror, depriving them of sleep and rest!'[8] Over these dread days and throughout the summer, the city was covered with posters – headed HYDROPHOBIA or MAD DOGS – that warned everyone to be on their guard and instructed that all dogs should be confined, and only be taken out muzzled and on a lead. The papers carried advertisements for popular remedies – specifics such as the Ormskirk Medicine and general remedies such as Morison's Universal Medicine – and medical men wrote in with their favourite means of halting the advance of the disease.

Medicine and hydrophobia

As the press reports in 1830 indicated, the summer months in previous years had also seen canine madness on the streets of English towns. Rabies canina, as it was then termed, was sporadic in the eighteenth century and was thought to have become more common after outbreaks in 1807 and 1808.[9] Thereafter, reports of cases in dogs and humans increased, and as a newly prevalent affliction it became the subject of many studies and treatises. Published accounts were mainly written by doctors and focused on hydrophobia, with asides on the condition in dogs. The leading British expert was the Manchester surgeon Samuel Bardsley, who was incidentally the first person to suggest eradication.[10]

Most doctors accepted that there were two conditions in humans: hydrophobia due to the inoculation of a virus in the saliva of a dog; and spurious hydrophobia – an hysterical condition brought on by fear of the consequences of a dog bite. The word hydrophobia derived from a mental state, a fear of water, which always included choking and the inability to swallow, and sometimes even anxiety at the sound of water being poured. Many patients were sent into convulsions by draughts of air across their throat and face. The most common and immediate problem facing doctors in 1830 was what to do with the many dog bite victims seeking treatment. Clinical experience was that the risk of developing hydrophobia from the bite of an obviously rabid dog was around 1 in 15, a figure which was constant throughout the century. Yet, in the Dog Days of 1830 there were no ordinary dog bites; the public sought medical attention for these injuries more than previously and doctors knowingly 'over-treated' to reassure and alleviate fears. There were two forms of 'treatment' – preventive and curative. The former aimed to stop the absorption and spread of the poison after the bite, the latter aimed to manage and counter the full-blown disease. Preventive treatments were almost always effective, while curative treatments always failed.

So, what preventive treatments were available? The first and most important aim was to remove the virus or halt its absorption, so victims were advised to suck their wound and to wash it vigorously and then to seek assistance from medical practitioners, principally surgeons or surgeon-apothecaries. Practitioners used two techniques – excision and cauterisation, and often both together. If the wound was on a fleshy part of the arm, leg, or trunk, then the surgeon would cut wide and deep to remove as much affected tissue as possible. If the wound was on the face, then surgeons endeavoured to destroy affected tissues by 'burning' with a hot iron or caustic chemical; lunar caustic (silver nitrate) was the most common agent. Dr Vaughan recommended that igniting gunpowder in the wound 'may have its uses' and in some cases amputation of a finger or part of a limb was undertaken.[11] William Lawrence wrote of the use of opium to induce sleep and slow the body's metabolism, while Magendie's inoculation of tepid water into a vein seems to have been aimed to dilute the poison.[12] The often lengthy time between the bite and the onset of symptoms led doctors to assume that the poison remained at the site of inoculation, possibly in nerves; hence, excision was tried for many days or weeks after the bite. Some doctors still resorted to cupping or bleeding, though these had lost favour generally and anyway it seemed that hydrophobia was a nervous rather than blood disease. That said, heroic interventions almost certainly had another purpose – avoiding spurious hydrophobia; treating a fearsome condition with formidable measures was one way of trying to reassure victims that hydrophobia was not inevitable and that anxiety was unnecessary.[13]

There were a great many remedies available to stop or counter the action of the poison for victims unable or unwilling to consult a surgeon. First, there were popular practices and superstitions, the best known being treating the wound with 'the hair of the dog that bit you' – treating like with like – which may have had resonances with homoeopathy. Related to this were the concoctions and remedies passed down in families and communities; many towns had a 'Mad Dog Man' who had advice on wound management and their own recipes for people and animals. Most local chemists and druggists had their own favourites or offered specifics such as the Ormskirk Cure, which as the name suggests was from Lancashire but had gained such a reputation that it could be found across the country.[14] Other popular remedies included chlouret of lime, inhaling the fumes of burning charcoal, alum, Armenian bole, calcined oyster shells, gall of the Dog Rose, liver and dried blood of a mad dog, cantharides, and immersion in sea water.[15] The London variation of this early form of aversion therapy involved finding a boat sailing down the Thames to Gravesend – the nearest point to the city where the salt content of the water was high enough for the required effect – and holding the victim under water until they nearly drowned.[16]

The situation regarding the therapies reported in the medical press was no less pluralist. In June 1830 the *London Medical Gazette* editorial lamented the great number of cures being canvassed by medical men, from bathing oneself (and one's dog) in sea water, to more drastic cures, such as amputation of the bitten limb. The author remarked that of 'the number of remedies which have been published within this fortnight, there is not one of them for any things indeed, they are calculated to do harm, by leading persons to place confidence where they will not find safety'.[17] This was followed by a statement that aimed to send a clear message about the severity and incurability of this disease, 'there is no cure and but a short period for prevention'. Finally, the editorial offered an alarming prognosis, 'We have known the disease come on though the part was cut out within half an hour after the receipt of the bite; the excision ought, therefore, to be instantaneous'. Those unfortunate enough to be bitten knew they might be living on borrowed time and this only heightened their anxiety and trepidation. Such assessments encouraged a search for methods of preventing rabies that aimed at controlling the recognised source of infection itself – dogs.

A modern perspective on rabies suggests that most preventive treatments of the 1820s and 1830s 'worked' because of the low infectivity of the virus and because most allegedly rabid dogs were probably not suffering from the disease. The same points were also made, albeit on different grounds, by contemporaries. Doctors pleaded repeatedly for biting dogs to be caught rather than killed so that the victim, the doctor, and the community could see if they had true rabies. This would soon be evident in the dog's worsening symptoms and death, and if it survived, victims might be spared radical wound treatment and weeks of anxiety. However, it was popularly believed that killing the dog denatured the poison; this may also have been the only way of getting hold of hair from the dog that bit you!

The initial symptoms of hydrophobia reported in 1830 were non-specific: a headache, pains in the chest and perhaps near the original bite, and general unease. Other characteristics were evident with hindsight, such as difficulty in breathing and an agitated mental state, so that sufferers became jumpy and sensitive to noises, light, and movements. Patients were sensitive to draughts on their faces and even the sight of water could provoke choking and fits. In time the patient would experience delusions, salivate profusely, become aggressive, and suffer paroxysms. Symptoms would last from 18 hours to 3 days; they would come and go, with periods of quiet followed by furious episodes. In most cases the patient died exhausted. All witnesses agreed that it was the worst of all possible deaths.

Medical descriptions often gave little sense of the mental agonies of the sufferer, the dangers endured by their carers, nor of the crisis that the whole episode constituted. To begin with, if hydrophobia was suspected doctors whispered to each other and tried not mention it at all for fear of inducing the spurious form. Sufferers were also usually taken to a private room to avoid frightening family or other patients, and to contain the expected pandemonium.[18] Doctors' accounts of the human disease were in part refracted through knowledge of the ferocious dogs that laboured under rabies. These narratives also reveal the social distance between the doctors and the lower classes. The human victims, nearly all reported to be male and of poor or low occupation, were subject to disturbing hallucinations that destroyed their sympathetic faculties, making them insensible to reason. Descriptions of rabid fits almost always included the patient's attempts to bite others or themselves, or the bedclothes, along with physical violence and verbal abuse. The following is a typical account:

About noon on the Wednesday I saw him. He approached me with clenched fists in a menacing attitude accompanied with a hysterical laugh, and a kind of howling noise, and great contortions of countenance. I naturally stepped back a little, when he composed himself, sat down, and told me I had no occasion to be frightened ... In about a minute he got up and rushed furiously across the room, then threw himself across the table like a person labouring under a violent fit of the colic.[19]

Here is another account, from May 1830, about which the doctor concluded, 'It could not be called aberration of intellect; it was ungovernable fury'.

At one of clock his irritability had augmented, the secretion of saliva was excessive, and he cast it about him in every direction. ... Another enema of guaco, as he refused to take it my mouth was proposed; this incensed him to a great degree, and it was necessary to threaten him with straps. His legs were secured, and on proceeding to confine his hands he, for the first time, showed a disposition to be vicious, for collecting saliva in his hands he discharged it in the face of the attendant. It was deemed necessary to secure him in a strait-jacket. ... No sooner, however, were his legs liberated, and one of his attendants had retired for the jacket, than, keeping others from him by spitting and throwing saliva upon them, he suddenly sprang out of bed, seized the large syringe which was filled with an injection of turpentine, and advanced against those present, spitting and discharging its contents at them. The gentlemen present thought it prudent to retire for a moment, and forgetting that the key was inside, shielded themselves behind the door.[20]

William Lawrence warned about mania and paranoia in a lecture that, indicating the public profile of the problem, was reprinted in *The Times* in July 1830.

The slightest causes will bring about a paroxysm, and the patient is pursued by a thousand fancies that intrude themselves upon the mind. He supposes he is holding converse with a great number of individuals; that persons are coming into the room to attack him; he fancies himself in danger, difficulty and distress. These thoughts come in rapid succession one after another, and keep the patient in a state of mental excitement.[21]

Such symptoms were typical of those shown by inmates of the new asylums; however, there is no evidence of asylum doctors showing any interest in hydrophobia. This would indicate that doctors saw it as akin to hallucinations and fits that often accompanied high fevers rather than madness as such.

Absence of all humane feelings to loved ones was customarily cited as diagnostic indicator of hydrophobia. As one account of a fit explained: 'the frenzy had reached a height at which he could not be soothed, even by his wife and sister, whom he severely attempted to injure by blows; their affection and firmness, however, eventually overcome the rage'.[22] The words of the rabid patient were often diagnostically important. In the same account the doctor recalled how,

> it might be well to mention that for two or three hours proceeding his dissolution he spoke nothing scarcely but bloodshed, frequently calling for his knife in order that he might annihilate those who had, in his perturbed imagination, been the cause of his sufferings, and amongst that number I was very particularly included.[23]

In nearly all descriptions, the human victim and the dog belonged in the same moral universe of barbarity and aggressiveness. When one doctor asked his patient why he barked and howled whilst in the fits, the young man explained that the dog that bit him stood at the foot of the bed, now with four heads and eager to recommence the attack.[24]

Doctors assumed that the minds of poor easily were emptied of morality and slipped back into animalistic rage, but stated that the minds of the middle and upper classes reacted differently. The class and gendered character of hydrophobia can be seen in accounts of those who possessed higher rank and position. Civilised minds, according to some medical experts, did not see troops of mad dogs, nor did they degrade themselves by lewd or verbal abuse. For instance, on some accounts the well-fortified mind of the Duke of Richmond, who died a well-publicised death from hydrophobia in Canada in 1819, did not give way to insensibility and fury.[25] In the final stages of his illness, the Duke allegedly sat down calmly to arrange some private affairs, and writing the most affectionate letters to those whom he loved. In another case, a doctor reported no fear in visiting the sickroom of the respectable Miss McClive.[26] In her final moments, she called for different members of her family, and spent her last breath imploring blessings on all of them.

The treatment of patients with developed hydrophobia was ameliorative. As was common with those suffering from insanity, recourse was made to restraint, both securing patients to beds and fitting strait-jackets. The newly favoured moral therapy was of little use to those who had descended into an animal state; hence, there were obvious echoes with earlier notions of the mad as people who had lost their humanity.[27] Symptoms were managed with sedatives, such as laudanum and opium, which because of the difficulties of getting patients to swallow were usually administered *per rectum*. Bleeding was often used, both to reduce the amount of poison in the body and to 'lower the system' – the depleted patient was also more manageable. In 1825, John Connolly reported the following:

> ten leeches were applied to the throat, and twenty five to the abdomen and thorax; and poultices of bread and milk, with the addition of laudanum, were applied to the leech bites in both situations [and] a blister was applied to the throat and another, the size of an ordinary pair of bellows, was applied to the ... thorax and the abdomen.[28]

As there was little to lose, doctors experimented. For example, in the summer of 1830, Elliotson reported using the newly promoted South American herb guaco, otherwise used to counter snake bites, while others turned to older methods such as cupping and bleeding.

There were also rumours that, faced with such horrific symptoms, doctors often resorted to euthanasia. Dr Thompson wrote in March 1828 that 'it is the belief that the government sanctions the smothering of patients labouring under Hydrophobia, between feather beds and that anybody may so serve another with impunity'.[29] Comments on this practice were widely aired, but unsurprisingly few specific cases were cited and most of these were from the eighteenth-or early nineteenth-centuries.[30] Smothering was said to rank alongside the burning of witches, though it took on new meaning in the 1820s in the context of body snatching – were doctors killing terminally hydrophobic patients to obtain cadavers? In 1833, James Bardsley may have been trying to reassure his readers when, in the context of the recent concern over grave robbing and dissection, he ended his discussion of smothering hydrophobic patients with the observation that doctors had little to learn from the bodies of victims, 'the frequent opportunities of inspection after death, now afforded by the zealously cultivated science of anatomy, has never cast any additional light on the subject'.[31]

Observations of the anxiety and fear in those bitten, not to mention the perception of a wider public panic, led many doctors to wonder whether hydrophobia was solely a product of an agitated mind. If mad dogs were allegedly everywhere on the streets and many people were being bitten, why did doctors see so very few cases of hydrophobia? This in turn raised other questions. How many 'mad dogs' were truly rabid? What happened in cases of people bitten who did not develop the disease? Did rabies have a lower infectivity in humans? Was hydrophobia actually the human form of rabies, or a wholly hysterical condition? The question – was the dog truly rabid – was often impossible to answer as the dog had been destroyed, but more often than not when it survived, the dog lived on and the incident proved to be a false alarm. To the question, why so many escaped the disease, most doctors replied that excision, cauterisation, and other physical measures were almost always effective when practised by competent surgeons, acting quickly and vigorously. To the third question – on low infectivity – doctors accepted this as fact, but explanations were contested. Some doctors argued that low infectivity was due to the fact that little or no poison was inoculated, with clothing wiping the poison from the biting teeth and absorbing saliva. Others argued that the effects of the poison depended on the mental and physical constitution of the individual bitten, the long latent period pointing to a struggle between the poison and the constitution. With the final question – on the relation between hydrophobia and hysteria – a minority of medical men maintained that all cases were in fact spurious, and that rabies in dogs did not transmit to humans. In other words, hydrophobia was always a form of mania or hysteria, brought on by anxiety and fear.

Doctors who considered rabies to be a contagious disease had to confront the fact that it was quite distinct from diseases that should have been analogous: small pox, cow pox, syphilis, plague, and snake bites. When these poisons were introduced into the body, the effects were evident within a limited and predictable time. With hydrophobia doctors found that the latent period could be anything from a few hours to a year; however, claims of 12-year lags between inoculation and symptoms were treated with scepticism. The delay led doctors to focus on the role played by the constitution, either in containing the poison in the bitten part or where the poison remained inert until victims were 'thrown out of health, frightened, till they catch cold, or something happens to disturb the constitution then it appears'.[32]

Those who argued that every human case was spurious hydrophobia were, of course, making the disease a wholly mental condition. The similarities with hysteria were in symptoms such as agitation, the production of saliva, the fear of swallowing, alternate fits of laughter, crying, and screaming. The surgeon, Benjamin Travers, explained that a girl under his care became so obsessed that she might become the subject of hydrophobia that 'although she is assured the dog is alive and well it has been necessary to produce the dog to convince the patient the dog is not mad'.[33] That this was a sure case of a 'disordered imagination' was discerned in how the sister of the patient had also lapsed into a state of low spirits from undue mental apprehension after her sibling's terrifying close encounter with a dog. Cases such as these demonstrated, at least in the opinion of this doctor, that human hydrophobia was not the result of any poison introduced into the host system, but merely melancholy, an often fatal result of panic and fear, and of the 'disordered state of the imagination'. Robert White, a doctor from Brighton, dismissed hydrophobia as a complete 'myth', arguing how 'it is the mental anticipation of the dreadful consequences that might arise'.[34] White recalled a half-starved, heart-broken weaver, who had worked himself 'off his legs' to keep his family solvent, and had become gripped with anxiety about an event 12 years earlier when 'a strange dog bit him'. White scornfully reported the case and observed how many Manchester medical men had believed that an event so longer ago could be anything other than spurious hydrophobia!

Veterinarians on canine madness

Veterinary practitioners were a very mixed group, from farriers to college-trained men.[35] Elite practice was dominated by the horse, a valuable capital asset on which owners would spend time and money. The highest status group were veterinary surgeons who looked after cavalry and other horses used by the military, and those who had retainers with the landed and wealthy. There was some demand for veterinarians on farms, especially for high-value pedigree breeding stock, but most diseased livestock tended to be summarily slaughtered in order to realise its economic value. There was little demand for the treatment of cats, dogs, and other pets; the repair costs were much higher than replacement costs, so most sick domestic animals were 'put down'.

Elite veterinarians, especially in London and Edinburgh, were aspiring to be a 'gentlemanly' profession. They were encouraged by the rising status of veterinarians in the Army and the possible demands for their services from improving agriculturalists. The shift in medicine away from humoral explanations of disease to those based on lesions and morbid anatomy had brought growing medical interest in comparative anatomy. Doctors had found it difficult to obtain human corpses for dissection, and so had turned to animals using homologies in teaching and research. Veterinarians welcomed this recognition of the similarities between human and animal medicine, yet also felt threatened because doctors also began to speak confidently about animal diseases. Throughout the early decades of the nineteenth century veterinarians suffered the strictures of medical men who, in their own struggle for status and wealth, were keen to distance themselves from the craft and trade associations of 'horse doctoring' and align themselves with the established gentlemanly professions of law and the clergy.[36]

London, with its mix of old and new money, was the only place in Britain where veterinarians could specialise in pet or small animal practice. Gainsborough portraits reveal the attachment that many aristocrats had to their pet dog or cat, and many also maintained packs of hounds on their country estates. In the 1810s and 1820s Delabere Blaine established a large dog practice in Nassau Street attached to the Middlesex Hospital, and became the country's first expert on canine pathology. In 1817 he published one of the first books on disease of dogs entitled *Canine Pathology*.[37] The volume covered all of the known conditions treated by veterinarians, detailed in alphabetical order, usually from one to four pages. The entry on rabies, listed under 'Madness', spanned 36 pages.[38] Blaine wrote that he devoted so much space to this condition because of its prevalence and the degree of misunderstanding that surrounded it, 'Perhaps hardly any other popular subject presents such a complete tissue of error as this', and admitted that it might be 'in vain to reason to prejudice'.[39] He was unhappy with all terms for the condition and stated that even 'Madness' was inappropriate as dogs hardly ever suffered 'total alienation of intellect'; they remembered people and places and would often obey commands. Yet, Blaine also suggested that the rabies poison changed the character of the dog, who developed a form of monomania – 'an instinctive desire to propagate the malady' evident in their roaming and biting. He set out in great detail the symptomatology, including the distinctive howl, and its morbid pathology. He dismissed all claims that the malady in the dog could be treated once developed, though he stated that excision was effective in animals as well humans. However, his main point was to show that 'Madness' never developed spontaneously, 'no dog breeds madness, that is, no dog becomes mad from any other cause whatever but his being bitten or inoculated by another dog'.[40] His point was that all notions on spontaneous generation due to the influence of the seasons, heat, drought, feeding, sexual frustration and ill-treatment were wrong. Blaine also argued against the identity of rabies and hydrophobia – they were distinct diseases in different species.

Blaine's pupil, then assistant, William Youatt, took over the Nassau Street infirmary and the position of Britain's authority on canine pathology. (See Figure 1.2.) He had to suffer the epithet of the 'dog doctor', but he had wide expertise and in 1828 began lectures at his infirmary on veterinary medicine. These were promoted as covering all species as against those at the Royal Veterinary College which dealt only with the horse. Against Coleman's vision, Youatt offered a more egalitarian profession that dealt with all God's creatures and served all citizens. His ideals were close to the reform agenda of radicals like Thomas Wakley, the editor of the *Lancet*, hence, the reprinting of many of Youatt's lectures in that journal.

In the summer of 1830 Youatt was the obvious person for the government to turn to for authoritative advice on rabies in London. In July, in response to the 'not unfounded alarm' that was gripping the capital, Youatt published a pamphlet entitled *On Canine Madness* that reprinted a series of articles that he had published in the *Veterinarian* in 1828 and 1829.[41] This was one of a number of publications that resulted from the furore of that summer, though the one to which other authors deferred.[42] In a similar vein to Blaine before him, Youatt saw his main task to be countering popular and professional errors. On symptoms in the dog he stated that,

> There is no dread of water; no spasm attending the effort to swallow; but a most extraordinary and unquenchable thirst. There is no fear excited in other dogs; no wondrous instinct warning them of danger. There is no peculiar smell; no running with the tail between the legs; except when weary and exhausted he is seeking his home; no pustules in or near the frenum of the tongue.[43]

Early symptoms were set out in great detail, as these were said to be so variable and non-specific that it was important to be comprehensive. Post-mortem findings were also itemised in great detail, as it was from this approach, in line with wider medical developments, that Youatt expected a fuller understanding of the condition to emerge. He gave special attention to the nervous system as he was convinced that the rabies poison affected the instinctive and involuntary actions of nerves in the throat and chest. He also maintained that most large quadrupeds were susceptible, with the different anatomies of the dog, cat, horse, cow, sheep, and humans producing different pathologies and symptoms. Thus, dogs showed unquenchable thirst because of their simple larynx, which was only able to bark and howl, and had low sensitivity to the

WILLIAM YOUATT, 1776–1847

Figure 1.2 William Youatt, no date

Source: From F. Smith, *Early History of Veterinary Literature*, London: Bailliére, 1930, Volume 3, Plate 5. Reproduced by permission of the Wellcome Library, London.

poison, whereas the greater complexity of the tones evident in human speech came from a more sensitive larynx that was more powerfully affected by the poison.

The key questions asked of veterinarians in the summer of 1830 were: what was the cause of the unusually high incidence and what could be done about it? Youatt's pamphlet directly answered these questions. Repeating Blaine's convictions, Youatt stated that 'rabies is produced by inoculation alone, and that the virus is confined to saliva'.[44] He dismissed spontaneous generation first by analogy, comparing rabies with small-pox and syphilis in human medicine, where it was accepted that the virus was always transmitted and was not repeatedly created anew. Indeed, the evidence was that rabies was perhaps the oldest of all known diseases as it seemed to have been described by the Egyptians. Next, he contested specific instances of spontaneous generation stating, 'in nine-teen cases out of twenty the inoculation can be proved. In almost every other case the possibility of it cannot be denied.'[45] The assumption of monomania was evident again as the dog was said to be 'bent on destruction'; indeed, Youatt argued that the poison transformed dogs, making them behave like criminals. He wrote in 1831 that 'it should never be forgotten, that there is in rabies a degree of treachery, a deeply laid plan to lull suspicion, and assuredly to accomplish mischief'.[46]

Rabies outbreaks were sporadic, so how was the power of the poison maintained in months and years between outbreaks? One answer offered was to extend the period of latency between inoculation and the development of the disease, which was already longer than any other known contagion; another was the possibility that some animals might harbour the poison without developing the disease. Youatt's faith in contagion was strengthened by the much cited experimental studies undertaken by Magendie and Breschet in Paris, where domestic animals inoculated with the saliva of a rabid dog went on to develop the disease.[47] Like most of his contemporaries, Youatt assumed that infectious and contagious diseases had many causes, some exciting and some pre-disposing, and that the first case in some epidemics arose spontaneously from certain configurations of the external physical environment and internal environment of the animal. Indeed, he accepted that distemper in dogs and glanders in horses were spontaneously generated; hence, his objections to the *de novo* origins of rabies were from experience and not principle.[48]

Youatt felt that most veterinarians shared his views on the extreme rarity of spontaneous origins, but he knew that many medical men and some veterinarians were against him.[49] In 1830, Thomas Pettigrew, a London physician, set out many of the 'origins' then current,

> The exciting causes of rabies have been supposed to be exceptional heat, putrid aliment, deprivation of venereal pleasures, combats, fit of exceptional passion, wounds secured during the rutting season, want of perspiration and the presence of what is called the moon under the tropics.[50]

No veterinarian now looked to lunar influences or to Sirius, the Dog Star that rises in July; however, many pointed to the heat of the Dog Days of summer and often linked this to the idea that rabies was brought on by dehydration.[51] Contemporaries assumed that dogs suffered in hot weather because of their coats and the fact that they did not sweat. The impervious qualities of dog skins were well known and before the Mackintosh were widely used for waterproof clothing. It was assumed that dogs relied on panting to cool themselves, which meant they kept their mouths open, often showing their teeth and salivating – so some speculated that rabies might be perverted, pathologised panting. The link with thirst was common across Europe and it was alleged that in Lisbon shopkeepers were required to provide buckets of water for dogs outside their shops.[52] The assumption was that dehydration led the blood to heat, or to putrefy and become poisonous. Thirsty dogs, it was said, would also lick up the 'putrid matter' (urine and faeces) from the street or their kennels, and that once in their body putrefaction would follow. And a poison created by thirst in the dog might produce the opposite in humans – an aversion to drinking.

On this notion, as in every other, veterinary opinion was divided. There was little evidence that rabies was more prevalent in hotter climes and there was experimental evidence from France that dogs left out in the midday sun for weeks did not develop rabies.[53] But the experience of rabies in successive years seemed to confirm the link between heat, thirst, and the disease, especially as in the summer months there were more dogs at large, towns were more crowded, and the streets more polluted and fetid.

The idea that rabies was produced by sexual excitement was championed in London by Henry Dewhurst, a Surgeon–Accoucheur.[54] The press and veterinary reports seem to show that the disease was found in male dogs more than bitches, and was more common in the latter when they were in season. A known symptom, though it was rarely mentioned in published works, was 'sexual excitement', which was manifest in priapism and attempts to mount other animals. As befits a surgeon, Dewhurst's views were based on anatomy. He noted that dogs, unlike humans, did not have seminal vesicles on the testis, which meant they had nowhere to store semen. Hence, if dogs were denied 'natural gratification' their semen would remain in the body, affecting the brain, either directly, or indirectly after putrefaction.[55] The fighting dog was also seen as a source of rabies as owners bred and trained their 'beasts' to attack and bite. Was it not possible that such dogs, which were owned by men who mistreated them, might develop the forms of mania seen in canine rabies? The association with foxes was also telling, as according to Dr Henry Thompson, the dog fox was 'an animal of salacious habits'.[56]

Youatt was clear that rabies could be controlled, 'if a species of quarantine could be established, and every dog were confined separately for seven months, the disease would be annihilated'.[57] Quarantine would require dogs to be kept indoors, or only to be taken out muzzled and on a lead, and also the removal of strays from the streets. However, he concluded regretfully that such a scheme could never be implemented.

> Then we must resort to other measures to lessen, if not terminate the devastations of this malady. Whence arises the evident increase of rabies? From the increasing demoralization of the country. From the lately adopted and cruel system of parochial government the peasantry of England has become degraded. The cottager is no longer enabled to support his family by honest labour; and the auxiliary pittance which the parish affords is doled out with niggardly a hand that he revolts at the acceptance of it. He tries other and fearful resources; he becomes a poacher – he is one of the organized gangs of nightly predators. To qualify himself for this, he provides himself with his dog, ostensibly to defend his little at all, but actually for the most nefarious purposes. ... In the large towns, within these few years, the dog pits, those nurseries of crime, have been established. The mechanic, the groom, the coachman, the apprentice, mingle there with the ruffian and the avowed thief. I will not speak of the barbarous deeds which are there perpetrated; but I will refer to the thousand instances, propagated nineteen times out of twenty, by the cur and the lurcher in the country, and the fighting dogs in town.[58]

Youatt presented rabies as a moral contagion, spreading out from sites of crime and depravity. On this reading, the cause of rabies was nothing less than The 'Condition of England' – the term that contemporaries coined for social and political crisis that reached a climax in the early

1830s. Concerns about the growing prevalence of rabies joined anxieties about the Swing Riots, the first Reform Bill, and anticipated arrival of cholera from Europe.

The rabid streets

Through the 1820s, *The Times* was inundated with letters from concerned citizens, most of whom were worried about the chaos on the street and wanted 'reform', a key word at this time. The view that canine madness had its origins in the way the poor treated its dogs and the way they lived on the street was widely held. One alarmed Londoner wrote to *The Times* in 1825 that 'whilst writing this, my trousers bear evidence of the danger of coming in contact with these quadrupeds, having accidentally stumbled over one in a very crowded thoroughfare, where it was impossible to see the animal'.[59] One fretful commentator recalled his uncomfortable closeness to curs, 'It is my lot, Sir, to live near a poor and crowded neighbourhood; and, notwithstanding the miserable spectacle which every where presents itself, I am shocked in seeing it aggravated by the filth, noise and contagion of innumerable cur dogs'.[60] Other writers contended that curs were the greatest danger to personal safety in the urban landscape, and some linked it directly to deviance and crime. Lancelot Baugh Allen, a former magistrate at New Union Hall in London, explained that the urban dog nuisance lay with 'those loose dogs, which follow persons, particularly idle and disorderly persons', and went on that 'a great many of the lower classes keep them as companions and a greater number of mischievous companions'.[61] Social reformers frequently assumed that the curs and their poor owners were indistinguishable: as Mr Allen explained, 'I live in Dulwich, and I am in the constant habit of riding out in all hours, and I am frequently set upon by mischievous persons who have dogs with them'. Such comments about the nature of urban dog population suggest that the main fear was less about rabies than a generalised anxiety about the moral status of the working class that was refracted through their dogs.

Canine madness moved from a problem to a crisis in the summer of 1830. A key sign was the cry of 'Mad Dog, Mad Dog' heard repeatedly across London and other towns. What normally followed was described in *The Times* on 3 June for an incident in Lambeth,

[T]he cry ... induced the passers by to retreat with precipitation into the adjoining shops and houses, the doors of which were instantly closed, so great and general is the fear of imbibing the horrid disease of hydrophobia. Three men, however, with several boys, with sticks, pursued the animal with praiseworthy intention, though a dangerous one, to kill the dog, which being pressed and harassed, ran amongst some children, who endeavoured to stop it, and bit one more or them. The dog was driven into a corner in Webber Street, from hence it was prevented from escaping by the exertions of several men with sticks. The animal by this time was foaming at the mouth and making a spring seized a young man by the right hand just above the little finger, but was instantly struck off, and by the aid of the broomsticks which pinned it to the ground it was finally destroyed.[62]

The resort to arms, to guns as well as sticks and pitchforks, further alarmed observers. However, working-class responses showed that while some enjoyed the sport it was important to remove the dangerous dogs from the street and to show no mercy. Reports of such incidents often told of heroic acts by individuals, but these were overwritten with anxieties about the mob, how rapidly it formed, and how dogs were dispatched in savage ways. Many commentators quoted Goldsmith's poem of 1766, which told the story of a dog that was cruelly killed by a mob only for it to emerge that it had never been rabid – a moral tale of the slaughter of the innocent.[63] Such a tale was told in Volume II of *Mirth and Morality*, published in 1834, where a 'falsely accused' mad dog was chased and killed, before it became known that it had been provoked into a fit by children tying a kettle to its tail. The dog escaping its tormenters was vividly depicted, with the moral message that 'We are too apt ... to believe every idle report, and to join in persecuting the miserable'.[64] (See Figure 1.3.)

Mad dog chases were sometimes portrayed as comic, as in the poem 'Hydrophobia' in W. H. Harrison's *The Humorist* in 1831 (see Figure 1.4), which was illustrated by a drawing entitled 'Kunophobia'. The illustration uses the mad dog chase to satirise popular discontent with the church.

'Mad dog! Mad dog!' See then he flies,
Of hearing no respecter;
If not one shoot or brain the brute,
He'll surely bite the Rector.

'Mad dog! Mad dog!' How fast he runs
With hundreds at his tail,
Each with some murd'rous weapon arm'd
Club, pitchfork, spade or flail.

THE MAD DOG.

Wouldst thou inflict no needless pain,
   And do thy fellow man no wrong,
With prudent, kindly care, restrain
   The poison of a slanderous tongue.

Figure 1.3 'The Mad Dog', 1834

Source: C. Bruce, *Mirth and Morality: A Collection of Original Tales*, London: T. Tegg and Son, 1834, 29–40, 39.

Boys join the throng, and, a spit in hand
The cook runs out; and see
The draper with his fowling-piece
Well-charged with No. 3.

He had done better to have fill'd
The deadly tube with chaff,
For he has miss'd the butcher's dog,
And hit the parson's *calf*.

Now, past the doctor, swift he scuds;
They'll never overtake him;

If was not stark mad before,
They've gone the way to make him.[65]

Beliefs about rabies were recurrently formed around fears of the break-
down of civility. In particular, such views expressed the conviction that
there was something indecent and dangerous about dogs that belonged
to the lower classes. One despondent observer at the end of the summer
bemoaned the fact that 'troops of dogs are still to be seem infesting
every quarter of the town, and no man can walk the streets on his ordinary business without the risk of being attacked by them'.[66] The dogs of
the poor were said to produce 'consumption, noise, filth and other
dreadful maladies, besides hydrophobia' and were associated with 'the
dissolute habits'.[67] The same correspondent continued his denigration
of curs: 'the dogs that scour our streets and put our lives in danger
belong, for the most part, to the lower orders of the gentlemen of the
fancy, and are kept for the laudable purpose of badger-baiting, dog-
fighting and duck-hunting on Sundays'. Another writer laid the blame
strongly at the door of barbarous individuals, 'the idle vicious characters

KUNOPHOBIA—THE CHURCH IN DANGER.

Figure 1.4 'Kunophobia – The Church in Danger', 1831

Source: W. H. Harrison, *The Humorist: A Companion for the Christmas Fireside*, London:
Ackermann, 1831, 259.

resorting every Sunday with their dogs to the fields round the metropolis – to bear and badger baiting in the week – getting drunk, and spending money at the expense of their famished families, to say nothing of their brutality and cruelty'.[68] These claims became the staple of speeches and arguments in support of stringent controls on dogs of the working class.

Some commentators drew attention to the morbid fascination of the working classes with cases of rabies and how this revealed their base appetites. In May 1830, *The Times* reported a dreadful scene where an allegedly hydrophobic dog turned on its master and his young daughter; the daughter's face was particularly badly mangled by the bite. The screams of the parties drew some people to the spot and a crowd then collected.[69] Sometimes, the authorities had to step in to restore public order. When Barrat, a resident of Drummond Crescent London, developed hydrophobia, he was heard howling and barking, and a crowd assembled round his house listening to the dreadful sounds.[70] The moral depravity of the spectators, who were likely, some weeks earlier, to have enjoyed the chase and the kill, was allegedly evident in their curiosity with the gruesome and terrifying scene.

In the specific context of the dogs of the working class and the streets they infested, medical men, veterinarians, and reformers pointed to four elements in their culture that produced and propagated canine madness: breed, neglect, exploitation, and mistreatment. The very character of the dogs, the specific breed or the 'non-breed' as with curs, was an important factor – nature and class ran together in contrasts between the breeds favoured by the rich and poor. As one worried commentator wrote, 'I have been frequently surprised at seeing the streets, especially thronged poor neighbourhood, loaded with what the fanciers term curs – that is dogs without breed'.[71] On one hand, the mastiff, bull, Newfoundland, greyhound, terrier, spaniel, pointer, sheep, and hound dog were breeds that were said to be useful, and,

> possess qualities peculiar to each breed, which qualities are preserved or destroyed as mixed together. ... The old contempt of that fellow, John Bull, may be crossed to advantage with a terrier; but the same dog with a spaniel would produce a mongrel, or cur, that is without any essential quality to recommend him, and which no fancier or judge would keep even for pay.[72]

Dog breeding symbolised prestige and leadership.[73] The lack of respect for breed amongst the lower classes had allegedly,

> filled our streets with hosts of ugly and useless curs, who remain in the undisturbed possession of their owners, not being worth stealing,

propagating, to the great scandal of the Society for the Suppression of Vice, their filthy species in the most indecent and unbounded manner in all our highways.[74]

It seemed to commentators that the only solution to the high incidence of hydrophobia was destruction of these low breeds of dogs: 'if these no-breed curs, which are at least as two hundred to one, were destroyed, there would be little fear of hydrophobia, every one being interested in and careful of a good and handsome dog.'[75] Cur dogs seemed to intensify the urban squalor and add to the general sense of disorder. This was the message delivered by one writer,

[Curs] suffered to roam at large, partaking of the most putrid food, not worth stealing, they infest our streets unmolested, creating noise, filth and general annoyance; not is this all, the peculiar sexual intercourse of the species renders them very dangerous. I have seen a whole gang of curs so strong excited, as to be little short of mad – so furious, indeed, as to change their very natures.[76]

Cur dogs were a law unto themselves: inbred, uncared for, dangerous, and furious.

Animal welfare reformers also argued that the working class neglected and exploited their dogs, and that many were in poor condition because their owners could not afford to feed them. Hence, they were turned on to the streets to scavenge for their own food, producing 'the painful sight of thousands of dogs, lost by their owners, wandering in a state of extreme exhaustion and starvation'.[77] Indeed, dog ownership was an indicator of working-class fecklessness and imprudence. The dog's was another mouth to feed when the family was already in poverty.[78] It also demonstrated ignorance on the part of owners, as many believed that 'Hydrophobia makes its appearance ... in dogs which exist in a state of confinement which are kept in towns and take little exercise'.[79] However, this comment was aimed at the pampered dogs of the middle and upper classes, as much as those of the poor. Reformers also claimed that dogs were made 'subservient to the avarice' of their owners in being made to pull dog carts.

Dogs are converted ... into beasts of burden; we hourly see them panting and gasping under trucks, their tongues protruded from their mouths, without the possibility of getting water, which is as necessary to cool them as to quench their thirst. By many persons an additional step or two is taken to promote rabies. Some keep the mouths of dogs firmly closed by a muzzle, during this labour, that there may be no opportunity of the animals cooling themselves by their mouths being open.[80]

A wag wrote to *The Times* under the pseudonym of 'Hector' complaining on behalf of his 'miserable brethren [who] are forced to submit to drawing trucks, barrow and other vehicles, driven at their utmost speed by a lazy fellow, till they sink to the earth from exhaustion; is not this, I ask, enough to cause madness?'[81]

Animal welfare reformers expressed greatest outrage at dog-fighting and the culture of the dog pits, which were said not only to be immoral, but also worked upon those who observed them to enfeeble and to coarsen their moral sensibility. Fighting pits were customarily represented as scenes of vice, a sort of carnival in which thieves, pickpockets, and prostitutes joined the dogs brought by 'butchers, brick makers and others who do not appear to live by honest industry'.[82] All those present became brutes. This was indicated in an account of events at the Westminster Pit in July 1830, 'groans of the bear, screams of the badger, the yelling of dogs, with shouts and blasphemies of the spectators, formed a discord of sounds that it was impossible to describe'.[83] William Drummond, a leading evangelical humanitarian, detailed the psychological principles at work in producing sympathy between the spectators and the cruel acts: 'To a mind of common sensibility, the pain of another creature seems to communicate itself by some sympathetic tie: in him it will at length cease to excite any emotion but pleasure'.[84] Even those with sensitive and educated feelings would succumb to the sights of these carnivals of cruelty; Drummond stated that the witness 'becomes the terror to his trembling domestics, a stern father, a savage husband and a tyrannical master'.

The pits were also a disaster for the fighting animals. In the case of dog fights, if the 'match' was proving to be a stalemate the dogs were separated 'by the men biting with their teeth the end of the tail of each'; the dogs were allowed to rest and then brought forward again. The fight ended when bets were given up, however, 'if the sums be of any importance to the owners, the contest is seldom decided till one or both of the dogs are bleeding, lacerated and almost dead'.[85] The link between fighting dogs and rabies was clear to many, a coroner concluded in June 1830 that 'this disease was propagated by fighting dogs; if an end was put to that practice, *rabies canina* would abate'.[86]

Dogs and criminality were manifest in other ways. The large number of strays in London had led to dog-stealing, whence carcases were sold to dog-skinners. A letter from Frances Maria Thompson to the *Voice of*

Humanity in 1830 elaborated on contemporary events. 'Dock-skinning has now become the regular trade of a class of men, of whom, in London, there are at least fifty leading characters, besides their spies and outposts.'[87] These men were hard to spot as they went incognito as 'plasterers, carpenters and carry the very tools in their hands, or they hawk about oranges, hardware and other articles of sale.' Thompson alleged that they caught their prey with nooses, enticed them with food, and that 'The periodical alarm of hydrophobia, and consequent order of the general destruction of dogs, produce to them an abundant harvest.'

While the increase in canine madness was usually blamed on the working class and their habits, other culprits were identified. Reformers trying to prevent cruelty to animals pointed to the ineffective collection of the Dog Tax, which had been introduced in 1796 by William Pitt and was set at five shillings, a not inconsiderable sum.[88] However, its collection was piecemeal and the working classes largely ignored it. One hope of reformers was that tighter implementation would reduce the number of dogs and ensure that all dogs were well cared for. In 1829 Youatt recommended that 'a tax be levied on every useless dog; and doubly or trebly heavier than on sporting dogs. Let no dog but by the shepherd's be exempted from the tax, except perhaps the truck dog'.[89] But Dr Henry Thompson blamed the Dog Tax and Pitt for the 'ravages of late' and 'the dread that everyone feels'.[90] He wrote to the Home Secretary that 'It is a malady particularly English' because nowhere else was there a tax that discouraged the owning of bitches. Nowhere else was there the same unnatural ratio of dogs to bitches, and nowhere else was there the same numbers of semen-filled, frustrated dogs. Thompson called for the abolition of the tax, and failing that its amendment to give lower rates for those who kept bitches or had their dog castrated.

The state and canine madness

If the state was going to act on rabies, contemporaries anticipated that it would be through immediate medical and veterinary police methods rather than changes to taxes. There were constant calls for coercive dog controls aimed at reshaping the dog-keeping habits of the working classes. Some local authorities had introduced police measures, but rabid curs were no respecters of administrative boundaries, which led to calls for national action. Colquhoun's treatise on London police matters, re-published in 1829, summed up the unsatisfactory state of the law as follows:

> When there is a rumour of mad dogs, the magistrate usually orders all dogs to be tied up; but this is a precaution for which, in the present

state of the law, the magistrate has not the authority to enforce. A person bit by a savage dog had a remedy of action; but in this he may fail, unless he can prove the owner had notice of the dog having bitten someone at least once before. Suffering a *mischievous* dog to go loose or unmuzzled is indictable as a nuisance; but it is too vague a description, and the remedy too difficult.[91]

Indeed, some people who were fined for letting loose their mischievous dog argued that there was no law to sanction such an attack on private property; the injured party had no right to destroy the dog nor did the police.

In the view of another commentator the paucity of controls on dogs could be related to the 'unhealthy' influence of nation's dominant political philosophy: liberalism. It was precisely the 'mistaken tenderness for the rights of the subject appears in this instance, Sir, to have reduced us to the state of a nation of fools, in standing with our hands tied behind us while subjected to the most appalling calamity that can befall a human being'.[92] Liberal ideas according to this understanding had simply left the public unprotected and exposed to the menace of mad dogs. In direct opposition to this was the viewpoint, drawing a direct parallel from the wider debate on Parliamentary reform and the extension of the franchise, that the muzzled dog was symbolic of the attempts to deny 'the people' their democratic rights. This is seen in William Heath's drawing, published on 1 July 1830, which featured a muzzled dog protesting against oppressive government actions on the grounds that Britain was a *'free country'* and that muzzles were 'a petty infringement on the liberty of the subject'. (See Figure 1.5.) The 'Notice to Dogs' is signed by Dogberry, almost certainly a reference to Alderman Mathew Wood, the Mayor of the City of London – Heath used the same name in a direct reference to Wood in another drawing that month.[93] The political uses of mad dogs was evident again in August, when King Charles X of France was portrayed fleeing his country in the wake of the Revolution; 'A dog (r.), shaved in the French manner, with the head of Charles X, and a battered kettle tied to its tail, flees before a crowd of men, women and children with tricolour flags and armed with pitchforks, & c.'[94]

Many proposals were put forward to reduce the numbers of street dogs. One anonymous petitioner to Robert Peel suggested an immediate and effective remedy, 'calling in the aid of the truncheons of the police on all loose dogs'.[96] Another more alarmist correspondent asked for regulations to be devised 'to insure perfect safety to human beings' and recommended 'the whole canine race to be destroyed'.[97] Another writer

Figure 1.5 W. Heath, 'Hydrophobia', July 1830. Courtesy of the National Library of Medicine.[95]

called for at least three quarters of the canine race to be 'made away with', if not it would be left to

> His Majesty's subjects to take the law into their own hands, and that bands be formed for the extinction of the whole canine race found at large: for surely, is it not better that 99 innocent dogs should be sacrificed than one mad one should escape to have the power of inflicting such calamity and such horrible feeling upon the human race.[98]

The crisis in late May and early June 1830 led the government to appoint a Committee to report on a Bill 'to prevent the spreading of canine madness'[99] The Committee began taking evidence promptly on 18 June and in all 15 witnesses were called: 2 veterinarians, 7 medical men, 3 magistrates, and 3 others – a policeman, a poisons expert, and a resident from the Strand.[100] The two veterinarians were sworn enemies and there was rivalry between the veterinarians and the medical men. Needless to say, there were also differences amongst the medical men, and between them and the magistrates and the police. And underlying all the proceedings were tensions between different types of professional and lay 'experts', with the Committee members having their own views. Witnesses were questioned on all aspects of the problem, with the issues of the nature of the disease and treatment occupying most time; and all of the uncertainties discussed earlier in this chapter were aired. If there was a consensus it was on the following: that hydrophobia was the human form of canine madness, that the disease was primarily spread by inoculation of a poison, that preventive treatments were effective, and that once hydrophobia had set in no remedy worked.

The first witness was William Youatt, confirming his position as the leading authority on the subject, though the other veterinarian was his arch enemy Edward Coleman, Principal of the Royal Veterinary College.[101] Youatt stated that canine madness was never spontaneously generated, while Coleman maintained that this was common; Youatt argued that the problem could be controlled by confining dogs; Coleman argued that this would produce the disease as they would become 'impregnated with an animal poison from the lungs, faeces, urine and skin'.[102] However, they both agreed that muzzling was desirable, as too was 'lessening the number of those breeds, the cur and the terrier ... and ... putting down assemblies of fighting dogs'.[103] Almost all witnesses concurred that the long-term solution was 'diminishing, very considerably, the number of dogs belonging to the poor, by taxation, or some other means'.[104]

All witnesses confirmed that there had been a recent increase in the incidence in canine madness amongst the dog population and that it had peaked in May and June 1830; however, some medical men doubted that there had been any increase in hydrophobia in humans. The number of cases seen in hospitals remained small, and leading London surgeons had doubts about many of the cases reported in medical journals and in newspapers. Their scepticism reflected the continuing uncertainties amongst medical and veterinary experts about the nature of rabies and hydrophobia, but it also indicated their opinion that this 'Era of Canine Madness' was largely a phenomenon of the streets and the popular imagination.[105]

Conclusion

'Hydrophobia is inflicting the horrors and tortures of the damned on the people.'[106] This was the view of an anonymous correspondent to the Home Secretary in June 1830 and many Londoners would have agreed. The crisis produced popular panics, middle-class outrage, veterinary and medical campaigning, ad hoc control measures by local magistrates, and a Parliamentary Bill to enable magistrates to issue notices ordering all dogs to be confined for a certain period. In the event, the Bill was never enacted, lay and professional apprehensions passed, with events soon overshadowed by the death of the King, the General Election and the continuing political crisis. If it was so quickly forgotten and brought no enduring change in the law or professional opinion, what was the significance of the summer of 1830?

The 'Era of Canine Madness' was about an urban crisis and a microcosm of the larger 'Condition of England question. Roaming, mischievous, biting, mad dogs were a metaphor of the rootless, uncouth, brutal, criminal classes, whose behaviour seemed to threaten to infect the whole working class. Mad dogs were no respecters of social class, geographical boundaries, or behavioural codes. Often people dealt with the problem themselves by chasing and killing mad dogs, but they also turned to the law and magistrates to deal with dog owners, and to the police to control disorder on the streets and to 'arrest' mad dogs. In turn, magistrates looked to the state for new and enhanced powers.

The 'Era of Canine Madness' brought dogs into the political arena in Britain. Parliament was reluctant to take on the question and unsurprisingly the matter was eventually dropped in 1831. However, the crisis highlighted the new sensibility towards man's best friend; it seems to have stimulated attempts to ban dog pits and it was no coincidence that

1830 saw the publication of *The Voice of Humanity*, the publication of the Association for Promoting Rational Humanity and Towards the Animal Creation. Brian Harrison has argued that the Society for the Protection of Animals, founded in 1824, switched in the 1830s from its initial concern with cruelty to horses and cattle, to urban animals, as evident in the 1835 Animal Cruelty Act.[107] We would suggest that the canine madness crisis was a key factor in this widening of animal welfare reform.

Third, and over the whole 1820s, canine madness was also used by those trying to reform the veterinary profession. Blaine and Youatt were able to advance their campaign against Coleman and the horse dominated teaching at the Royal Veterinary College by showing the importance of domestic animal practice, not only to individuals but also to the nation. Youatt gained in stature from his position as the leading metropolitan expert on the disease, and from association with medical men. He advanced the claims of veterinary medicine in utilitarian terms, arguing that humans had

[Taken animals] from their native plains, and coerced and confined them, and too often exacted their labour with reckless cruelty and enacted on them, by our absurd practices, or too frequent and disgraceful brutality, numerous diseases, and a premature death. I might content myself, however, with these undeniable facts, that they are susceptible of pleasure and of pain; that we [veterinarians] can never be unprofitably or dishonestly employed while we are increasing the former and diminishing the latter, and that, while thus employed, we are in the strictest sense accomplishing the grand object of medical science.[108]

Youatt was seeking equivalence with medicine not only on the basis of comparative pathology but also on the basis of new grounds, namely, that veterinarians' patients had 'intellectual, 'ay, and even moral qualities'. The only difference between human and veterinary medicine was 'value of the patient' in economic terms.[109]

It is a moot point whether the standing of medical men was enhanced by their responses to canine madness and hydrophobia. Their speculations on the disease in dogs were readily undermined by the greater experience and authority of Youatt, who nonetheless sought alliances with doctors against his veterinarian enemies. The weakness of the medical response was commented upon by John Murray who wrote in the summer of 1830 that 'While we contemplate, in the rapid advancement of science, such bright discoveries and noble achievements, Hydrophobia remains a blot on her shield and escutcheon – the unconquered Goliath of disease – and one victim after another falls'.[110] The summer of 1830 did not bring closure on the many medical questions; indeed, if anything it highlighted uncertainties. For example, in his entry in Forbes's *Cyclopaedia of Practical Medicine* in 1833, James Barclay felt he had to begin by reviewing the evidence of the 'very existence' of hydrophobia, and then took 35 pages to cover every nuance of opinion, and then said nothing about the control of the disease.[111]

The lack of certainty amongst professional experts was a problem, especially as the 'Era of Canine Madness' was seen by some commentators as an opportunity finally to end the battle between the forces of reason and unreason – between science and the nonsense that had been and was so much a part of popular belief. An anonymous medical man writing in the *Westminster Review* in October 1830 reflected on the recent crisis in a long historical perspective.

> A witch! The plague! and a mad dog! Behold the Trinity which long held the dominion of fear over mankind. The days of the first of the trio are at an end; scarcely can anyone be found to pay homage. The plague, though no trifle, is viewed with less horror, because its nature is better understood, and it may be, at all events, avoided by not entering the fatal locality. A mad dog still exercises a fearful influence over almost all the thinking as well as the unthinking portions of society.[112]

He went on to hope that 'the star of its ascendancy may be on the decline' and looked forward to the cooperative and collaborative efforts of medical men ending the anguish, apprehension, and unfounded fears surrounding hydrophobia.

2
Rabies at Bay: 'The Dog Days', 1831–1863

Between 1830 and 1860 rabies was overshadowed by wider problems in pubic health and public order. Public health became dominated by zymotic diseases, especially epidemic diseases such as cholera, typhoid fever, scarlet fever, and typhus, which contemporaries understood to be mostly generated by filth and malign conditions, and to be spread by aerial miasmas.[1] Hydrophobia was classed as a zymotic disease, though like syphilis spread by contact contagion. The fight against epidemics was led by so-called sanitarians who emphasised aerial transmission of disease poisons and focused on the physical environment, principally the water supply, sewage, and nuisance removal; in their view few important diseases seemed to be spread by direct contagion.[2] Public order became political order, first in the reform crisis of 1830–1832, then in the introduction of the New Poor Law, and then the Chartist campaigns, which seemed to some contemporaries to threaten revolution.

The reported incidence of rabies and hydrophobia dropped and both were experienced as sporadic events that affected individuals, families, and neighbourhoods, rather than threatening towns and the wider social order as they had in 1830. However, the diseases did not disappear from the public consciousness. The unique combination of fear, excitement, comedy, mystery, and tragedy that was remembered, and still periodically experienced, ensured their profile was maintained in elite and popular culture, and in medicine and veterinary medicine.

Although hydrophobia deaths were reported throughout the year, public anxieties and local authority control measures concentrated on the Dog Days, the five weeks between 8 July and 11 August, when many towns were covered with posters warning of 'Hydrophobia' and 'Mad

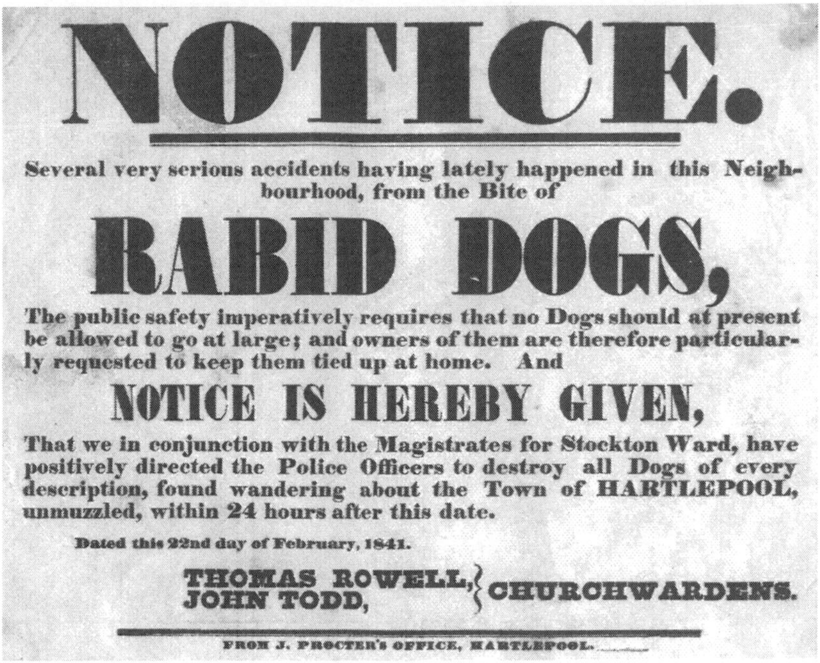

NOTICE.

Several very serious accidents having lately happened in this Neighbourhood, from the Bite of

RABID DOGS,

The public safety imperatively requires that no Dogs should at present be allowed to go at large; and owners of them are therefore particularly requested to keep them tied up at home.   And

NOTICE IS HEREBY GIVEN,

That we in conjunction with the Magistrates for Stockton Ward, have positively directed the Police Officers to destroy all Dogs of every description, found wandering about the Town of HARTLEPOOL, unmuzzled, within 24 hours after this date.

Dated this 22nd day of February, 1841.

THOMAS ROWELL, } CHURCHWARDENS.
JOHN TODD,

FROM J. PROCTER'S OFFICE, HARTLEPOOL.

Figure 2.1   'Notice: Rabid Dogs', 1841

Source: PortCities Hartlepool, Robert Wood Collection, ID: 2556, John Proctor, 22 February 1841. Courtesy of PortCities, Hartlepool.

Dogs', or requiring muzzling and confinement of dogs. (Figure 2.1.) Cries of 'Mad Dog, Mad Dog' continued to cause public panics, the killing of strays, and controls on dogs. Reports of local mad dog incidents and deaths were syndicated nationally, with headlines that typically used adjectives such as 'shocking', 'dreadful', 'melancholy', 'distressing', 'lamentable', and 'horrible'.[3] Narratives continued in a familiar melodramatic vein, as two examples from the early 1840s demonstrate. The author of letter to the *Animal's Friend* in 1840 wrote:

> I feel it is my duty to offer a few simple remarks to the destruction of two dogs, reported to be mad in Hyde Park two or three days ago, and hope you will join in your usual strain of humanity of denouncing the shameless slander so often passed by human beings devoid of reason in order to indulge their spite on poor dumb animals by beating their brains out.[4]

In the following year, Richard Beal, a surgeon observed:

> The senseless cry of mad dogs so often raised in Town does produce the most wanton cruelty to these faithful animals. If a cry was raised by a mad man! Mad man! Do you think the Mayors of Towns would require all men found loose in the street to be shot?[5]

Such comments indicate the changing sensibility towards dogs in the 1830s and 1840s, some individuals and groups spoke up for dog, giving them human characteristics such that they could be slandered and compared to mad men. Although most mad dogs were said to be strays or curs, some were portrayed as innocents who were brutally murdered by irrational crowds. The change in sensibility was, unsurprisingly, most evident in the writings of animal welfare reformers who claimed that the 'faithful dog' had more humanity than 'ignorant people' who were 'devoid of reason', and who 'maddened by drink and rage' put harmless dogs to such awful deaths.

Rabies, the street, and animal cruelty

In 1831 Alderman Mathew Wood tried to introduce Canine Madness Bills to Parliament, each containing stronger measures, for example, allowing the destruction of stray dogs not just their confinement. He failed to find support and gave up the struggle somewhat aggrieved because the House of Commons would not take his proposals seriously. Wood was an ineffective Parliamentary performer, whose radical liberal views were disliked by the Tory government of Wellington and Peel, which felt dogs were something that Parliament should not be legislating on.[6] He tried over 40 times to introduce a Bill but failed every time. Events on the streets were also against him. The summer of 1831 saw far fewer reports of mad dog incidents and deaths from hydrophobia, and the numbers declined further over the next decade. Indeed, when national mortality figures were first compiled from the late 1830s, they were on a downward trajectory and from 1843 to 1846. (See Graph 2.1.) Although the figures show no deaths in the mid-1840s, newspapers regularly reported such deaths. The discrepancy is almost certainly due to the unevenness and unreliability of death certification.

Official figures show no deaths in London between 1846 and 1853; hence, the increase that began in 1846 was mainly in Lancashire and West Riding of Yorkshire. However, the pattern of reported hydrophobia deaths should not be taken as a proxy for the incidence of rabies in the dog population, or the number of mad dog incidents for that matter.

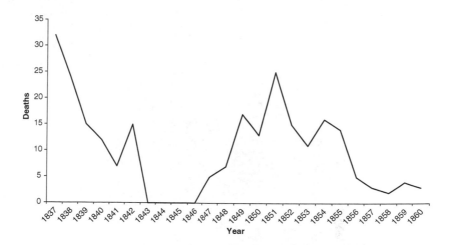

Graph 2.1   Hydrophobia deaths in England and Wales, 1837–1860

For deaths between 1838 and 1860, see G. Fleming, *The Nature and Treatment of Rabies and Hydrophobia*, London: Baillière, Tindall & Cox, 1878, 122. The figure for 1837 is from: *Lancet*, 1841, 36: 87.

Most dogs suffering from rabies would not have bitten a human, so mad dog incidents say more about the volatility of the street than the disease state of the hounded animal.

Awareness of the threat of rabies was maintained in the 1830s through popular memory of the 'Era of Canine Madness', the number and type of dogs on the street and, in the summer months, by the numerous warning posters. And after a mad dog incident, by the experience or reports of round-ups of strays, court cases, fines, and slaughtering. For example, in Manchester in the 1830s stray dogs were shot on the street, but this was abandoned when the dog-man's aim went awry and he hit a young boy in the leg.[7] The shotgun was replaced by the hatchet, until complaints that the agonies of half-dead dogs left on the street were offending the ladies of the town led to strays being rounded up, taken to dog pounds, and poisoned with strychnine.

The character of the street and the varied lives of dogs in the 1830s were illustrated by the drawing for July 1836 in George Cruikshank's *Comic Almanack*.[8] (See Figure 2.2.) Written and published at the end of 1835, Cruikshank drew an imagined streetscape of working-class London for the coming summer.

Unsurprisingly the scene is overrun with dogs of all shapes and sizes, engaged in all manner of activities, and associated with all forms of

Figure 2.2 *'The Dog Days', 1836*

Source: G. Cruikshank, *The Comic Almanack: First Series, 1835–43*, London: John Camden Hotten, 1871, 59.

humanity. Dogs are working as beasts of burden pulling dog carts; they are dancing, performing, and fighting; they are being bought and sold, shopped for, and exercised. On the right, leaving the scene is a pack of hounds in the back of a carriage driven by two grooms in top hats. There is only one 'Mad Dog' poster, but the number of other posters indicates the importance of the poster in public information and advertising. From the mid-1830s to the mid-1850s, the *Comic Almanack* regularly looked forward to the Dog Days, mostly in rhyme, as in 1837 when it was linked with St Swithin's Day.

> Two potent elements combine
> To rule the month together,
> St. Swithin gives us showers of rain,
> The mad dogs, *biting* weather.
> And if you get a dubious gripe
> From Pincher, Snap or Toby,
> The good saint's bucket comes right
> To test the Hydro-phoby.[9]

In 1853, the publication was still satirising the 'Dog Days' predicting that:

> July will be a very hot month. Several cases of hydrophobia will occur. In each instance the dog will be killed as soon as he has bitten a sufficient number of people to amount to a conviction. The theory of prevention, by muzzling or chaining up, will be suggested by many people, but will continue to be disregarded, as entirely opposed to the spirit of the British Constitution. A terrible act of injustice will be committed. A very sensible dog indeed will be killed as mad – for refusing to drink a drop of Thames water.[10]

The idea of the dog as a 'criminal' and its summary execution suggests that popular attitudes to 'Man's Best Friend' and rabies had changed little from the 1820s; however, also present was the idea of the innocent and sensible animal, whose owner, and perhaps the dog itself, had rights.

In the early 1830s the Society for the Protection of Animals (SPCA) was joined in its reform efforts by the Association for Promoting Rational Humanity towards the Animal Creation (APRHAC).[11] Both campaigned to extend the 1822 Animal Cruelty Act (the Martin Act), which formally only covered cattle, to include domestic pets and animal combat, especially bull-baiting, cockfighting, and dog pits. Drawing together humanitarian and political ideals, reformers sought to align cruelly treated animals with suffering humanity. For example, they increasingly pointed to 'dumb animals'; the link between having no voice and no vote was made obvious, but there were also notions of childlike innocence and of the exploited slave. Rabies was central to attempts to prevent cruelty to dogs. Its threat was mobilised from the premise that, if rabies could be generated by ill-treatment, then greater humanity towards dogs would be of physical as well as moral benefit of humankind. In other words, outbreaks of rabies were not 'Acts of God' or random natural events, but could be traced to human ignorance and folly. The specific danger of spontaneously generated disease was raised with respect to dog-fighting, dog carts, and to the nature and breeding of the dogs of the poor.

Dog pits

After the 1832 election, the push for animal welfare reform in Parliament was led by Joseph Pease, the newly elected MP for South Durham.[12] His father had promoted the Stockton and Darlington railway, and the family had extensive industrial and mining interests; they were Quakers and had been active in antislavery organisations. Joseph Pease promoted education and social reform, as well as animal welfare; of course, in his mind all these areas were linked. The first fruit of the campaign was the appointment of a Select Committee to look at consolidating laws relating to the ill-treatment of animals. This took evidence in the early summer of 1832, in the midst of the cholera crisis, and reported in August.[13] To the old concerns with 'the sufferings of dumb animals, and the demoralisation of the people' was added a new unease over the cruelties perpetrated by the urban poor.[14] The limitations of the Martin Act meant that action against animal cruelty in urban areas had to be pursued through nuisances legislation, which in the early nineteenth century was principally aimed at pollution, fly-tipping, and noxious trades. The main animal nuisances were those associated with livestock, particularly driving cattle through city streets, urban dairies, and slaughtering.[15] However, animal welfare reformers had other targets, principally bull, bear and badger baiting and cock fighting, now showing that their aims were about both the morality of the poor and the welfare of animals.

Much of the evidence taken by the Select Committee was on dog fights and the culture of the dog pit; indeed, the two were indistinguishable. William Youatt was called and confirmed 'the infallible connection between fighting dogs and the alteration of character for the worse'. He stated that he personally knew of a servant-man who, after attending a pit, became 'a worthless character and a beggar', and a young man in 'the high grade of society' who became 'a very different character from what he would otherwise have been'. He condemned pits on three grounds: 'inasmuch as they encourage cruelty in the first instance, inasmuch as they tend to increase rabies in the second instance, and inasmuch as they tend to bring people together of bad character'.[16] When asked about dog pits and rabies, Youatt restated his view that rabies was a contagious disease and never arose spontaneously; his point was that dog pits were spreaders of the disease rather than its source. Moreover, because the fighting dogs had been trained to be aggressive, they would propagate the disease ten times more than a pet dog.

Owners of dog pits argued that they had reformed their business and that it had diminished in scale, such that there were now only four pits remaining in London. They also stated that the pits had become more respectable. John Roach, proprietor of the Smithfield pit, claimed that his fights were attended by aristocrats and the middle classes, that there was no crime and that neighbours did not complain.[17] Indeed, Roach claimed that there was no cruelty; he said that he always made fair fights by matching dogs by weight, and that good dogs were valuable assets, and hence, they were well fed, exercised, and used for breeding. William Hemmings, a dog doctor and dog-trader, also claimed that fights were not cruel as a dog could not be made to fight, though questioning revealed that part of his business was attending to injured dogs. Both Roach and Hemmings denied that fighting dogs were more likely to develop rabies. Instead, Roach associated rabies with the common cur – '[It was] those dogs that run about in the streets loose and pick things off the road, and run after the females, picking up all kinds of filth and dirt that go mad'.[18] Hemming too claimed that he had seen many street curs go mad, but he had never seen fighting dogs turn. One questioner suggested that hydrophobia was more prevalent in England, compared to other many countries because of dog fighting; his answer was that rabies was not as common as many supposed and that most 'mad dog' incidents were due to popular panics. This was yet another example of the claim made repeatedly by dog lovers and those in the dog business that rabies was nowhere near as prevalent as commonly believed.

The Select Committee also explored the matter of breeds, a key question being, Were certain breeds – the most popular were the bull-dog and terrier – more likely to produce and propagate rabies than others? Youatt said not, yet again having to resist the notion of spontaneous origins. He argued that small spaniels and bull-dogs were equally susceptible, what mattered was how often they went out and whether they were protected by their owners. Hence, if valuable to their owners, fighting dogs would be well kept and protected, and hence might be less likely to catch or spread rabies. However, the bull-dog could not be so readily dismissed. It was an emblem of liberty and closely bound to patriotism and the national qualities conveyed in the figure of John Bull. For working-class owners it symbolised resistance to authority, so those who sought to confine and muzzle these dogs were also seen to be trying to suppress such national qualities as courage, strength, and independence. The ambiguity of the bull-dog was further evident when it was adopted by the Chartists.

Dog-fighting was outlawed in the 1835 Act but continued in secret, pursued by the special constables of animal welfare groups.[19] Occasionally prosecutions were brought as in April 1846 when four men were tried at Bow Street Court.[20] The police had been alerted and an Inspector, several constables, and officers of the Animals Friend Society tried to infiltrate the gathering. They were spotted and a cry of 'Police!' led to attempts to clear the scene. However, the police managed to detain 88 people and brought four to trial. The Inspector who led the raid described the grim scene he found.

> A chandelier with candles was suspended over the pit, the floor and boarding of which were besmeared with blood. They found 14 bull and bull-terrier dogs ... some of them bleeding and foaming at the mouth. There were also a live badger, in a box used for badger drawing, a dead badger (but quite warm), some dog skins, buckets of water with sponges and the usual apparatus employed on such occasions. The prisoner Mansfield, whose arms were stripped and covered with blood, at once admitted he was the proprietor of the room, but declared he and his associates were only engaged in rat-killing.[21]

The defence complained that the police had been partial in bringing only poor men to the bar and discharging the 'noblemen and gentlemen' who were present and had encouraged the event. There was much concern about the class of those at the pit and the judge reassured the court that 'no gentleman would sanction ... such disgraceful and brutal exhibitions'. Mansfield was found guilty, and he immediately paid his fine and was then reported as 'proceeding homeward accompanied by a large fierce dog ... and was followed by hundreds of the lowest characters'.

There was, of course, a direct link in the minds of reformers between the 'lowest characters', their curs and rabies; a correspondent to the *Times* in 1837 wrote, 'If only well bred, and therefore valuable, dogs were allowed to exist, hydrophobia would soon become unknown'.[22]

Dog carts

While animal reformers were unanimous about banning dog pits, when they moved on to reform the use of dogs as beasts of burden they were divided. Some found dog carts as cruel and uncivilised as bull-baiting and dog-fighting, but others argued that it was possible to treat a draught dog humanely, or that banning dog carts would lead to greater neglect and cruelty when dogs lost their economic value. The campaign against dog carts took reformers into the working-class economy and their home, as most owners and drivers worked from home. At the Annual Meeting of the SPCA in 1835, Lord Mahon reflected on the limits of Acts of Parliament 'to make people humane' and gave domestic dogs as an example of animals beyond the reach of legislation.[23] Interestingly, he argued that the Society should avoid laws that led to 'any inquisition into private life', rather they would seek to appeal to public feeling and that removing cruelty from the streets would provide a lesson for the public to take back into their homes. Also, the attack on dog carts was part of a larger of concern about cruelty to all beasts of burden, notably horses and oxen being whipped, poorly fed and watered, and run into the ground. Reformers bemoaned the growing pace of urban and rural life, not least due to the competition between the railway and animal-drawn vehicles.[24] Dog carts attracted reformers because they were such a visible form of cruelty. An address published by the SPCA in 1836, while not condemning dog carts outright, drew attention to the loads and pace, and painted the carter as a grotesque figure:

> A strong Dog, of some particular breed, may no doubt be of much use in assisting an industrious and humane man gain a livelihood – but how is this latitude abused by the present practise of letting out for hire, by the day, poor little miserable curs to be beaten or starved, or driven at the pace of a fleet horse, with a monster riding in a cart behind.[25]

Over the 1830s, the SPCA changed from seeking to reform the use of dog carts to an outright ban.[26] (Figure 2.3.)

Figure 2.3 *'Dog Days – Legislation going to the Dogs'*, 1844

Source: G, Cruikshank, *The Comic Almanack: Second Series, 1844–1853*, London: John Camden Hotten, 1871, 34.

Reformers strengthened their claims for a ban by associating dog carts with the production of rabies. A letter in the APRHAC journal *Voice of Humanity* in 1830 lamented that, 'we hourly see them panting and gasping under trucks, their tongue protruded from their mouths, without the possibility of getting water, which is as necessary to cool them as to quench their thirst'.[27] The writer continued that some carters 'even keep the mouths of dogs firmly closed by a muzzle during this labour, that there may be no opportunity of the animals cooling themselves by their mouths being open; and others ride in the cart and urge the generous dogs by the lash of their whip'.[28] It was at this point of exhaustion, generated by lazy, dissolute drivers, that reformers claimed the dogs became feverish and rabid. These themes became the subject of humour in a poem in Cruikshank's *Comic Almanack* for 1837 entitled *How to Make a Mad Dog*, which also showed how a mad dog literally turned the streets upside down.

> Tie a dog that is little, and one that is large,
> To a truck or a barrow as big as a barge;
> Their mouths girded tight with a rugged old cord (or
> They'll put out their tongues) by the magistrate's order;
> So you save'em the trouble of feeding, I think,
> Or the loss of your time by their stopping to drink.
> Lend'em out, 'tis a neighbourly duty, of course,
> And mind they've a load that would stagger a horse.
> If you've nothing to draw, why, yourselves let 'em carry (sons
> Of she dogs!), or else they'll be drawing compare-sons
> With a stick or a kick machete gallop away,
> And smoke though the streets in a piping hot day,
> Where Mac Adam is spreading his pebbles about,
> And they'll pick up their feet all the quicker, no doubt;
> More than all, don't allow them their noses to wet; – it
> Will keep 'em alert by "wish they may get it."
> All pleasures must end:- when they drop head and tail,
> With their muzzles all froth, like a tankard of ale,
> Turn 'em loose in the road with a whoop and a hollo,
> And get all the thieves and the blackguards to follow.
> It's a precious good lark for the neighbours, you'll find,
> With the mad dogs before and the sad dogs behind;
> And you'll never be molested, rely on my word,
> If you keep'em from biting a Bishop or Lord.[29]

Part of the objection to dog carts was the claim that cart owners hired their dogs daily from dog farmers who kept packs of 20–30 animals; hence, carters would have no feelings for their dogs and no interest in their well-being.[30]

Dog carts were first banned, not through animal cruelty legislation but in the Metropolitan Police Act, 1839, which only applied within a 15 mile radius of Charing Cross.[31] This legislation once again followed deliberations by a Select Committee to which Youatt and others spoke on the connection between cruelty and rabies.[32] First, Anthony White, a surgeon at Westminster Hospital, gave evidence on the horrors of hydrophobia and its untreatability. Next, William Sewell reiterated the views of his mentor William Coleman that hydrophobia could be brought on spontaneously by ill-treatment. Needless to say, in his evidence Youatt disagreed, arguing only that cart dogs were more likely to spread rabies because ill-treatment made them aggressive. He recommended the regulation of dog carts through licences rather than an outright ban, linking this to the widely held view that outlawing dog-fighting had merely been pushed the practice underground. It is telling that dog carts were banned in the context of police measures which aimed to bring order to the street, not animal cruelty as such, though the two issues were, of course, inseparable. Rabies was also a factor, as animal welfare campaigners continued to suggest that cruelty could produce the disease.

The 1839 Act gave the police authority to confine and destroy dogs suffering from rabies, or suspected thereof, and to fine owners ignoring confinement orders. This was a success for the SPCA in its aim of reforming a situation where, 'herds of Dogs should infect every avenue in London – being both useless, and at the same time dangerous in many points of view'.[33] As well as being a danger to public safety and order, ferocious dogs were nuisances, and a threat to health from soiling the streets to spreading rabies. Indeed, the sporadic, but nonetheless regular mad dog incidents seemed to demonstrate the barbarism that lay just beneath the surface of modern street life.

There were immediate attempts by the SPCA to extend the ban to the whole country. Parliamentary Bills were introduced in 1840, 1841, and 1843, but all failed. These efforts were helped by the fall in the number of hydrophobia deaths in London and its absence after 1843, which reformers attributed to the impact of the 1839 Act. So, why should not the rest of the country benefit from the controls enjoyed in London? The situation elsewhere was similar to that in London in the 1830s. Towns experienced sporadic mad dog incidents, local magistrates would introduce confinement or muzzling orders, or both, and occasionally a victim would die a horrible death. However, rabies was not always linked to dog carts, rather it was the problem of strays.[34] For example, in 1845 in Manchester, 1,844 dogs were 'found at large' in the town: 1,498 owners were fined, 34 cases were dropped, and 273 dogs were destroyed.[35] Such measures were justified 'on account of apprehension of canine madness', but there was public division about their necessity. On the one hand, there were many who feared rabies and saw every dog as potentially rabid; on the other side were those, most of who seem to have been dog owners, who maintained that the danger was exaggerated, being fed by the appearance of warning notices every summer, as Mayors abused their powers.[36]

The rise in the number of mad dog incidents and hydrophobia deaths in the late 1840s and early 1850s prompted reformers to try again to have dog carts banned from the whole country. In 1854, Dr Webster confirmed that the absence of rabies from London was due to the banning of dog carts, in which 'hydrophobia said to be often produced'.[37] As well as preventing rabies, reformers also claimed that a ban would halt cruelty, protect other road users, and reduce crime.[38] New Bills were brought forward in 1851 and 1852, with success finally coming in 1854 with the passage of what contemporaries termed the Dog Cart Bill, though formally the ban was included with amendments to the 1849 Prevention of Cruelty to Animals Act.[39] There was a debate in Parliament in 1854 over this measure, with the use of dogs as 'beasts of burden' finding many supporters, who drew a calculus that compared the suffering of dogs against the hardship of the poor.[40] Critics of the proposed ban attacked the idea that the fear of rabies should triumph over the daily bread of the poor.[41] They also argued that animal welfare reformers were wrong about the cruel practices of dog owners, often quoting Youatt, they claimed, that the practice was not necessarily cruel. After all, it was not in the interests of owners to starve or overwork their dogs. One correspondent to the *Times*, 'Cynicus', argued that a dog 'is as well fitted for draught as the horse, the ass, or the ox'.[42] He went on to explain that when a dog was tied to a horse-cart, 'the moment the horse begins to move, the dog of his own accord strains every nerve, and seems to rejoice in the fancied success of his efforts'. The Dog Cart Bill was said by critics to have been driven through by 'ostentatious humanity-mongers' intent on unleashing unnecessary hardship upon the poor with little consideration of the political implications. 'Cynicus' worried about the double standards between the poor and the rich, recommending that reformers in Parliament 'should pitch their flight at a higher quarry. ... Let [them] introduce a bill to forbid the hunting of the noble stag, [and] the timorous hare'.[43]

Such suggestions expressed an unease that the reforms would be taken to be unfair and hypocritical: the new law was not congruent with English principles of justice and fair play, and there could be consequences if the poor recognised this. Opponents had long suggested that the measure would lead to greater cruelty as 'animals would be consigned by their owners a miserable death or driven loose upon the public, to become for a time a general nuisance.'[44] A Parson confessed that it was no longer safe to go and conduct Bible readings amongst the poor as the streets had become too dangerous. He described a vast stray population that lurked in the dark and shot out at you snapping and barking. This made his profession an ordeal; he wrote, 'I have visited rooms filled with cholera, fever and small pox, and do not complain; in my profession I am prepared to meet with it, but to encounter furious dogs I think a just ground of complaint'.[45]

The 1854 Act also gave local authorities the powers to kill suspected mad dogs and to introduce confinement and muzzling orders. These measures introduced a national standard to meet a situation where different authorities were using different powers to implement different measures. The intentions behind the Act were lampooned by opponents, for example, a Cruikshank cartoon published in 1853 shows an animal charity providing muzzles for 'dogs in humble circumstances'. Interestingly, however, the muzzle wearing dogs were humanised, walking on two legs and in conversation. (Figure 2.4.)

In the capital, despite the legislation, there continued to be complaints about the number and behaviour of 'London Dogs'. They were a problem on the street and in the home. A correspondent to the *Times* wrote in 1854 that:

I am compelled to walk to the city every day, and have suffered much annoyance for the multitude of dogs which prowl about the streets. A few days since I counted from the Obelisk in St George's-road, to Blackfriars-bridge, no less than 47 dogs, and I can affirm that other parts of the highways of London south of the Thames are equally thronged with brutes that might go mad and do a great deal of mischief.[46]

The year before a parish surgeon reported:

In my daily walks among my poor patients I meet with many perils and face 'the various ills that flesh is heir to,' with the accompanying risk of contagion. This I expect and contract for; but I do not contract for lacerated legs and hydrophobia, to which I am daily liable, and which constitute my grievance, and not mine only, but that of clergyman, scripture reader, and district visitor, who, I am happy to bear testimony, and unremitting in their attentions to the sick and unfortunate poor, and in like manner worried by this pest. Of 16 houses I have this morning visited, there were dogs in nine, and I have been bitten by one, making the fourth time in little more than as many weeks.[47]

Both correspondents pointed to the irresponsibility of the poor. However, a more complex picture of the working class and their dogs was emerging. The new discourses were no longer about sporting or working dogs, but the domestic pet and its male owner. Various versions of the irresponsible man were offered, at one pole were those men who indulged their dog to the extent that they 'will starve their wives and

Figure 2.4 'A distinguished philanthropist will institute a charity for the providing of dogs in humble circumstance with muzzles'.
Source: G. Cruikshank, *The Comic Almanack: Second Series, 1844–1853*, London: John Camden Totten, 1871, 397.

children to keep curs', while at the other pole were men who were socially irresponsible turning 'into the streets a parcel of snarling and half-rabid curs, to endanger the lives of his neighbours'.[48]

As legislation and appeals to humane feeling was unlikely to change behaviour of the poor in the home, reformers began to call for the rigorous implementation of the Dog Tax, which they claimed would restrict ownership to the more prudent and respectable working class and also raise revenue for the government.[49] In fact, one of the original reasons for the introduction of the Dog Tax had been to try and limit the number of dogs on the streets of towns. Calls for reform of the tax were most common in the Dogs Days of July and August, with the complaint that the temporary measures introduced then were 'too late to be of much avail'.[50] The biggest problem was that payment was largely voluntary. It was a complex tax to administer as the amount paid depended on breed and there were many exemptions; also the rates were relatively high and paid in a lump sum. Hence, there was large scale avoidance, except by the responsible middle and upper classes whose dogs were assumed to be already well cared for.

Brutes and the Brontës[51]

The power of mad dogs and hydrophobia in the popular imagination was evident in contemporary fiction, and this was especially the case in Emily Bronte's *Wuthering Heights* published in 1847 and Charlotte Bronte's *Shirley* published in 1849.[52] In *Wuthering Heights* neither rabies nor hydrophobia are directly mentioned, yet they are implicit in the text as wild dogs and rabid symptoms loom large throughout. The novel was set in an environment on the brink of barbarism, with the moors and their inhabitants depicted as beastly and untamed. The human relationships were refracted through mid-Victorian conceptions of human and animal relationships, with dogs frequently used to characterise nature of human relationships. One way in which the otherness of Wuthering Heights was presented was by the presence of untamed dogs, swarms of common curs that run amok in the house, undomesticated, untamed, and unfettered by the habits of pet-keeping. Wuthering Heights was portrayed like the street, full of dogs and liable to spontaneous pandemonium. The principal narrator, Mr Lockwood, was bitten by one of these wild dogs on his first visit. Heathcliff had warned him, 'you'd better let the dog alone ... She's not accustomed to be spoiled – not kept for a pet'.[53] However, when Lockwood went to the cellar he was surrounded by a 'tempest of worrying and yelping' and things soon became volatile.

> [I was] not anxious to come in contact with their fangs, I sat still – but imagining they would scarcely understand tacit insults, I unfortunately indulged in winking and making faces at the trio, and some turn of my physiognomy so irritated madam, that she suddenly broke into a fury, and leapt on my knees. I flung her back, and hastened to interpose the table between us. The proceeding roused the whole hive. Half-a-dozen four footed friends, of various sizes, and ages, issued from hidden dens to the common centre. I felt my heels and coat – laps peculiar subjects of assault, and, parrying off the larger combatants, as effectually as I could, with the poker, I was constrained to demand, aloud assistance from some of the household in re-establishing the peace.[54]

Not long after this encounter, Lockwood on opening a door was shocked when 'two hairy monsters flew at my throat, bearing me down and extinguishing the light'.[55]

Lockwood describes Heathcliff as a brooding figure, a fierce monster who literally growls at others – filled with rage and vehemence. The reader learns that Heathcliff was found on the streets of Liverpool – homeless and fatherless – like a common cur 'as good as dumb' and 'half dead with fatigue'. He was treated badly by Earnshaw's son Hindley, who thrashed him and called him a 'dog', and while Heathcliff took the treatment coolly it is suggested that his sympathetic faculties were damaged. Indeed, Hindley treated many people like a dog, including his son whom he tried to harden after not living up to expectations. He calls to him, 'Unnatural cub, come hither! I'll teach thee to impose on a good-hearted, deluded father – Now, don't you think the lad would be handsome cropped. It makes a dog fiercer, and I love something fierce – Get me a scissors – something fierce and trim'.[56]

When Cathy first visits the neighbouring Lintons, a more respectable family, she was bitten by their bull-dog Skulker in a scene replete with rabid meanings.

> 'Run, Heathcliff, run!' she whispered. 'They have let the bull-dog loose, and he holds me!' The devil had seized her ankle, Nelly: I heard his abominable snorting. She did not yell out – no! she would have scorned to do it, if she had been spitted on the horns of a mad cow. I

did, though: I vociferated curses enough to annihilate any fiend in Christendom; and I got a stone and thrust it between his jaws, and tried with all my might to cram it down his throat. A beast of a servant came up with a lantern, at last, shouting – "Keep fast, Skulker, keep fast!" He changed his note, however, when he saw Skulker's game. The dog was throttled off; his huge, purple tongue hanging half a foot out of his mouth, and his pendent lips streaming with bloody slaver.[57]

Cathy became close to the son of the family, Edgar Linton, and starts to don the garbs of civility held by the gentry class. He stays with her at Thrushcross Grange until her ankle improved and during this time her character seemingly changed. When she returned to Wuthering Heights 'her manners improved' and 'instead of a wild, hatless little savage jumping into the house, and rushing to squeeze us all breathless, there lighted from a handsome black pony a very dignified women'.[58]

Later in the story, when Cathy was prevented from meeting with Heathcliff, she develops a fatal brain fever. 'Tossing about, she increased her feverish bewilderment to madness, and tore the pillow with her teeth, then raising herself up all burning, desired that I would open the window'.[59] Then as the wild wind blew in from the moors, she convulsed more violently and vehemently. Violent reactions to currents of air on to the face were, of course, a common diagnostic sign of hydrophobia. In her dying moments, Heathcliff was described as having been overcome, with Lockwood reporting that 'he gnashed at me, and foamed like a mad dog, and gathered her to him with greedy jealously'.[60]

Charlotte Bronte's novel *Shirley*, published in 1849, was set in the 1820s and 1830s in the wool towns of the West Riding against the backdrop of Luddite activity and the social impact of the new industrial economy. Its two central characters were Shirley Keeldar, a woman of independent means and strong character, and her retiring cousin Caroline Helstone. The main plot concerned their relations with two brothers, Robert and Louis Moore, which was resolved when Shirley married Louis and Carole married Robert. Much of the novel was about the search of Shirley and Caroline for useful roles in society and how contemporary attitudes and conventions denied this to women. It is noteworthy that the gender politics of the novel are sometimes expressed through dogs.[61] Throughout the novel, Shirley flouts conventional gender rules, often donning masculine roles, even titles. Shirley was not dependent on a surrogate male: she was a wealthy heiress who had her own house and a mill. Like a Squire, she despised lap dogs that were usually associated with women of her social status and position. Rather, she had huge mastiff – Tartar – that never left her side and made her seem as if she was a Lord of the Manor. However, over the course of the novel the character of Shirley became outwardly more passive as she settles for marriage, though her internal strength remains. This was shown to readers in her reactions to being bitten by a dog she feared rabid. As with the great majority of such victims, Shirley did not develop the disease and the plot moved on.

This incident has not been given much significance in modern readings of the novel and few literary scholars have discussed it; however, against the profile of rabies and hydrophobia set out in this book, it takes on a new significance. There was no public intimation of her worries about hydrophobia, merely that family and friends noticed that Shirley seemed unwell. Though she became weaker and her behaviour stranger, she continued to deny any illness all. In front of others, Shirley put on a strong face, demonstrated reserve and strength of character by refusing to reveal her fears. However, her anxieties began to slip out. She was found waking in a storm and when comforted replied: 'the best thing that could happen to me would be take a good cold and fever, and so pass off like other Christians'. Louis Moore puzzled, 'That Shirley thinks she is going to die.'[62]

Eventually Louis confronted Shirley and her secret was revealed. She showed him her arm, which bore a rather deep, though healed, indentation, something between a burn and a cut. Shirley said that 'small as it is, it has taken my sleep away and made me nervous, thin and foolish; because, on account of that little mark, I am obliged to look forward to a possibility that has its terrors'. She revealed that she had been bitten by a local pointer bitch called Phoebe, which she had seen running with her head down and tongue hanging out. The bitch was bruised all over and had been ill-used, and Shirley had given her some water and dinner. However, the dog was agitated and snatched at her arm when she tried to pat her head. Shirley went on:

> She bit ... so as to draw blood, then ran panting on. Directly after, Mr Wynne's Keeper came up, carrying a gun. He asked if I had seen a dog. In reply Mr Wynne recommended that Shirley had better chain up Tartar. ... 'Phoebe is raging mad'. On returning to her house, Shirley took an Italian iron from the fire, and applied the light scarlet glowing tip to my arm; 'I bored it well in'; it cauterised the little wound.[63]

Shirley had lived in a state of fear. She imagined the effects of the virus, anticipating an indefinitely threatening, dreadful doom. In her final hours, she expected to become raving mad. Louis suggested that Shirley was 'very nervous and womanish', especially as feared that doctors might suffocate her between mattresses. She implored Louis to protect her.

> You know, in case the worst I have feared should happen, they will smother me. You need not smile: they will – they always do. My uncle will be full of horror, weakness, precipitation; and that is the only expedient which will suggest itself to him. Nobody in the house will be self-possessed but you: now promise to befriend me – to keep Mr. Sympson away from me – not to let Henry come near, lest I should hurt him. Mind – mind that you take care of yourself, too: but I shall not injure you, I know I shall not. Lock the chamber-door against the surgeons – turn them out, if they get in. Let neither the young nor the old MacTurk lay a finger on me; nor Mr. Greaves, their colleague; and, lastly, if I give trouble, with your own hand administer to me a strong narcotic: such a sure dose of laudanum as shall leave no mistake. Promise to do this.[64]

Shirley's imaginings refracted some of the major items in the novel – imprisonment, submission, impotence, and passivity, though she retained enough of her old self to have plans to end her life by suicide.

Shirley's health was eventually re-established as Louis tried to 'relieve her from every nervous apprehension', but in the process she lost her mental strength: 'every fear seemed to have taken wing: her heart became as lightsome, her manner as careless, as those of a little child, that, thoughtless of its own life or death, trusts all responsibility to its parents'. Enquiries by Louis about the dog suggested that it was not mad, rather it had been ill-treated and had become aggressive. However, its owners maintained that the dog had been rabid and that 'if [Shirley's] was not a clear case of hydrophobia, there was no such disease'. Louis Moore chose to ignore these views; first, because they showed typical ignorance of the disease, and second, so as only to give Shirley encouraging news.

Medicine, 1831–1860

In the early 1830s medical understanding of hydrophobia stabilised around three assertions: that the disease was always caused by the inoculation of a poison or virus from an infected animal; preventive treatment by excision and cauterisation was highly effective; but that once developed, the disease was untreatable, producing terrible suffering and inevitable death. These views were clearly set out in James Bardsley's article on 'Hydrophobia' in *The Cyclopaedia of Practical Medicine* in 1833.[65] The author was the son of Samuel Bardsley of Manchester, and in succeeding years the article was much cited as an authoritative summary of medical opinion. James Bardsley set out a detailed history of views and endeavoured to simplify the classifications then in use. He attributed much of the uncertainty to the confusion of secondary with primary symptoms, and to the highly variable incubation period, which he attributed to the interaction between the virus and the constitution of the person bitten. After a long discussion of spontaneous origins, Bardsley concluded that in the absence of positive evidence, inoculation of a virus-poison was the only proven cause. On curative treatments, he dismissed all specifics, commenting that there were probably more nostrums for hydrophobia than any other disease, though he recommended measures to ameliorate suffering and moderate symptoms.[66] A similar and longer work in the same vein was written by James Pettigrew in 1834, but the manuscript was never published.[67] One reason for this was the declining incidence of hydrophobia, and while sporadic incidents continued to be reported in the medical journals, the disease lost the medical profile it had previously enjoyed.

In these irregular reports and speculations, new perspectives were articulated and the medical consensus began to break down. First, reports of new treatments continued to be published; indeed, it seems that every new form of therapeutics was tried, including the new opiates, chloroform and ether, and new specifics from abroad, for example, South American Woorali (Curare) and Guaco. Indeed, in 1834 Youatt acknowledged that on very rare occasions sufferers might recover if the dose of poison had been tiny or became exhausted.[68] Second, there was more interest in spurious or hysterical hydrophobia. In his lectures on diseases of the nervous system in 1833, John Elliotson argued that the 'true disease' and the 'fancied disease' were quite different in terms of symptoms, but was clear that people could torment themselves into a degree of insanity.[69] Many doctors termed these cases 'pseudo-hydrophobia' and suggested that such cases produced most, if not all, of the claimed recoveries for the disease.[70] In 1836, Francis Eagle, a London surgeon, argued that there was an identity between hydrophobia and hysteria, in their nature, symptoms, morbid appearances and causes, and that this was due to 'physical deviation of the sexual system'.[71]

Accepting the 'individual and solitary' (i.e. spontaneous) origin of disease poisons, he drew a parallel that just as 'the venereal poison originates "under our noses" daily', so did 'the poison of hydrophobia in the very cat or dog by our fireside'.[72]

At the end of the decade, Youatt publicly criticised the editor of the new edition of *Hooper's Medical Dictionary* for the statement: 'Hydrophobia: Of the cause of this peculiar distemper in dogs nothing certain is known; that it originates spontaneously in them is now the general opinion'.[73] Youatt claimed that 19 out of 20 veterinarians denied the spontaneous origin of rabies and the same opinion prevailed in medicine, citing the evidence presented to the 1830 Committee. However, by the end of the decade there was greater acceptance that not every case of rabies came from contagion. This was part of a wider acceptance of anticontagionism that was evident with cholera, which vied with hydrophobia as the most feared disease of the age. Between the 1831–1832 cholera outbreak and that in 1848–1849, public health precautions shifted from focusing on the movement of people and goods as carriers of the disease, to measures that were aimed at predisposing causes, which endeavoured to make the physical and social environment less favourable for disease to develop and spread. This was for two main reasons: first, that many zymotic diseases were spread principally via the aerial miasmas; and second, that these diseases could arise spontaneously, either *de novo* in appropriate conditions or from the modification of other diseases.[74] Instances were normally rare, but with certain predisposing conditions they might become common. Such views were evident at a discussion on hydrophobia at the Royal Medical and Chirurgical Society in April 1852.[75] Mr Brodhurst was clear that 'the disease occasionally arose spontaneously' and that diet and climate were key factors. Hence, in Britain 'a dog ... allowed a diet principally of animal matter during the summer ... would almost certainly die hydrophobic', while he suggested that extreme cold and privation favoured the disease in wolves.[76]

The rising incidence of rabies and hydrophobia in the early 1850s produced a new spate of case-reports and speculations, with opinions on how the disease arose and spread as divided as ever. This was exemplified in an article published in the *Edinburgh Medical Journal* in January 1855 by Lindsay Kemp, in which the diseases were re-imagined through the lens of the recent cholera epidemic.[77] Kemp concluded that rabies in dogs was a non-contagious zymotic disease:

[It] would appear in every respect to resemble those epidemics that affect man, and depend upon temporary atmospherical causes. It breaks out suddenly, prevails extensively, and suddenly disappears, not to be seen again for years. During these intervals of absence no exposure to heat, or thirst, or inoculation with saliva of previously rabid dogs can produce it.[78]

In humans he concluded that hydrophobia was either 'traumatic tetanus' – an affliction that sometimes followed deep cuts – or a hysterical condition, 'produced by moral causes', to which persons with 'a "mobile" nervous habit or system' were susceptible. Kemp admitted that there was no consensus with medicine, with many doctors continuing to hold strict contagionist views. Another indication of uncertainties was that in 1855 and 1856 John Netten Radcliffe, who was at this time medical superintendent of the Hospital for the Paralysed and Epileptic in Queen Square, London, but who later became an influential Medical Officer of Health, published a series of articles on hydrophobia in the *Lancet*.[79] He hoped that by reviewing over 100 cases in the literature, for which he had to go back to the eighteenth century to compile, he would throw light on this 'obscure disease'. In all he published five articles over a period of 14 months, but the series ended abruptly as the 'to be continued' promise at the end of the last instalment was never fulfilled. It is perhaps wholly appropriate that such a review at this date ended without a conclusion.

Veterinary medicine, 1831–1860

William Youatt continued to be the leading veterinary authority on rabies through the 1830s. He included the most comprehensive available account of the disease in his lectures at the University of London, and he was regularly called upon by Parliament to give evidence to committees of enquiry.[80] He also used the *Veterinarian* to advance his views and published articles and letters from supporters, and occasionally from those with whom he disagreed.[81] Along with his mentor Delabere Blaine, he continued to maintain that rabies only ever arose from inoculation and that confinement, along with muzzling, could control the incidence of the disease. He also argued that the disease could be treated preventively if dog bites were cauterised or excised; indeed, given that he claimed to have been bitten thousands of times by mad dogs, he seemed to be living testimony of its efficacy! On the other hand, Youatt's claim to have survived so many normally fatal bites may have damaged his credibility.

Youatt's standing as a professional gentleman was undermined in 1832 when he was declared bankrupt. He had to sell his Nassau Street

Hospital and turned to writing to help maintain his income, publishing books on the nature and diseases of cattle (1834), sheep (1837), and the horse (1841) for the Society for the Diffusion of Useful Knowledge. His volume on the dog did not appear until 1845 by which time his strict contagionist views on rabies were increasingly against the tide of medical and veterinary opinion.[82] Doctors and veterinarians who held anti-contagionist views did not deny that diseases were spread by contact, only that in many instances they spread without contact and that poisons could arise spontaneously. The most significant disagreements were over policy, whether quarantines, isolation, and other state authorised police measures should be implemented. In the case of rabies, Coleman, Sewell, and others argued that the confinement and muzzling dogs would produce rabies because irritation of the nervous system would lead to madness. In 1842, William Dick, the founder and head of the Dick Veterinary School in Edinburgh, wrote in his *Manual of Veterinary Science* that rabies in the dog was an inflammatory condition and that biting was 'an accidental concomitant' and not the cause of the condition.[83] He also contended that hydrophobia in humans was non-contagious, being the result of 'melancholy fear and a disordered state of the imagination'.[84] In 1844, William Percivall, a leading military veterinarian, argued that rabies in the dog was like glanders in the horse, the result of exposure, in confined conditions, to their own urine and faeces.[85] Inoculation experiments undertaken at European, mainly French, veterinary schools, were regularly reported in the *Veterinarian*. The results almost all showed that inoculations of saliva, blood, and other tissues from rabid animals sometimes produced the disease and sometimes did not. Needless to say, anticontagionists took this as compelling evidence that factors other than direct contagion were involved.

Youatt's *The Dog*, published just before his suicide in 1847, covered dog breeds, their qualities, cruelty issues, as well as veterinary medicine topics. It was representative of a new type of dog publication, in which diseases and their treatment were set in the context of an appreciation of breed, character, and advice on management. For example, William Martin's *The History of the Dog: its origin, physical and moral characteristics, and its principal varieties* (1845), H. D. Richardson's *The Dog; Its Origin, Natural History and Varieties. With Directions for Its General Management*, (1851) and Edward Mayhew's *Dogs: Their Management. Being a New Plan of Treating the Animal, Based Upon a Consideration of His Natural Temperament* (1854).[86] There were parallels between this elevation of breeds and their geographical and hierarchical classifications with the emergence of race as an important category in the human sciences.[87] Mayhew's volume was the most successful; it was richly illustrated and became an essential manual, principally for responsible middle class owners on how to treat different breeds to achieve a well-behaved and healthy animal.[88] With rabies, he typically tried to evoke sympathy for the suffering animal, writing that 'Dreadful as hydrophobia may be to the human being, rabies is worse in the dog'.[89] He argued that rabies was a nervous affliction, even suggesting that it might be produced spontaneously by 'exciting the nervous irritability of the dog'.[90] Indeed, this had to be the case, as the first instance, however long ago, could not have been from contagion.[91] In making such claims, the new dog experts were, of course, echoing the anticontagionism of the wider veterinary and medical professions.[92]

Throughout the 1850s little was published on rabies in veterinary journals, either in the form of news items or substantive articles. The number of hydrophobia deaths suggests that there was little rabies around and professional work and communication continued to be dominated by large animals, above all the horse. However, the *Veterinarian* reported unusual outbreaks of rabies; for example, in 1856 there were stories of 'A Herd of Rabid Deer' and 'A Flock of Rabid Sheep'.[93] These were mostly news items reprinted from local newspapers which gave melodramatic descriptions of symptoms, for example, in the latter incident, 'The ewes that went into a rabid state, trotted backwards and forwards by the sides of the fold and repeatedly bit at the hurdles, and tore mouthfuls of wool out of each other, foamed at the mouth, &c'.[94]

The allegedly rabid horse attracted most attention, being depicted as a terrible, menacing, destructive spectacle: 'the animal kicks in the most violent manner, often attempts to seize and bite other horses and attendants, and will level everything to the ground before him – himself sweating, snorting and foaming amid the ruins'.[95] Such descriptions raised questions about the relationship between the horse and humanity as one contemporary wrote:

> Although the horse is a noble animal, there is neither the motive for, nor the capacity of that attachment the dog feels for his master. Therefore, under the influence of this disease, he abandons himself to all its dreadful excitement. There is also in the horse, whose affection for his owner is so easily and so often transferred, a degree of treachery which we rarely see in the noble and more intelligent dog.[96]

Seemingly, dogs were now seen as more deserving of humane treatment as they had human qualities such as affection and intelligence.

In 1859, John Henry Walsh, alias 'Stonehenge', published *The Dog, in Health and Disease*, a book that went through many editions and became a standard work for dog owners and fanciers for many decades.[97] Rabies was discussed as an inflammation and the author's attitude was fatalistic. He stated that, 'At present there appears to be little or no control over this horrible complaint', so he avoided speculation on causes and communication and instead focused on the earlier recognition of symptoms.[98] Owners were told to look out for changes in the temper and behaviour of their dog, and 'a peculiarly hollow howl', and not to obstruct a suspicious animal, otherwise they would excite its fury. Walsh was clear that certain breeds were more liable to nervous diseases, especially those bred for particular characteristics, for example, courage with the bull-dog, speed in the greyhound, and alertness in the pointer. Against this, he wrote, 'the cur, the common sheep-dog, and c. seldom suffer any disease whatever'.[99] The important point, however, was his assumption of an intimacy between dogs and families, so that it was imperative owners be aware of the early, insidious symptoms. Dogs and owners had become closer in two senses: first, physical proximity where the dog lived indoors, had freedom of the home and lived close to the family; and second, emotional familiarity, where the owner was aware of the dogs moods and emotions.

Conclusion

In 1861, George Henry Lewes published an article on 'Mad Dogs' in *Blackwood's Magazine*.[100] Lewes is now best known as the partner of the novelist George Eliot (Marian Evans), but in the mid-Victorian period he was an influential writer and critic, who numbered among his friends Charles Dickens, William Makepeace Thackray, and Anthony Trollope.[101] In the mid-1850s he had taken to writing essays on scientific subjects and he later became a leading light in the Physiological Society; indeed, Marian Evans endowed a scholarship in his name that still supports junior researchers.[102] Lewes took the need for the public understanding of science very seriously and fell out with Dickens complaining that the spontaneous combustion episode in *Bleak House* was 'beyond the limits of acceptable fiction and gave credence to a scientific impossibility'.[103] His essay on 'Mad Dogs' was similarly severe on public misunderstandings of science, beginning with an attack on popular 'vulgar errors', such as that mad dogs were phobic to water, that they always foamed at the mouth and showed 'furious ferocity', and that rabies was common in the heat of the 'dog days' of summer.[104] Drawing mainly on the work of the French veterinarian Sanson, Lewes set out an account of a disease spread wholly by inoculation, with different manifestations in man and animals, and about which there was much need for public education. He endorsed Youatt's plan for the extirpation of rabies by strictly enforcing the confinement of every dog and cat in the land for eight months.

For a great many of Victorians, strict controls on dogs, even to eradicate the dread rabies, were unacceptable and unnecessary. Severe penalties, especially against the poor, risked provoking or even hardening the most dangerous classes. Debates over dog controls were set in the broader frame of uncertainties over the nature of the disease, which allowed protagonists to produce different assessments of the threat. A critical question was, What proportion of cases originated spontaneously? The paradox with rabies was that animal welfare groups and elite doctors and veterinarians both wanted state intervention; however, they sought the same goal by different means. Welfare reformers sought to remove the moral causes of rabies in cruelty and neglect, while veterinarians and doctors were content just to prevent the transmission of its poison.

Advocates of control at source by halting cruelty to dogs argued that a critical failing of modern society was its savage treatment of animals, especially by the impoverished classes. The concern with the moral consequences of certain attitudes towards animals was part of a growing discomfort with all forms of violence and acts of cruelty in the early Victorian period. The campaigns of humanitarian reformers were fought across a number of issues, from capital punishment to slavery, and aimed to restructure society upon the principles of human feeling and compassion. A straightforward social message underwrote the preoccupation with this kind of reform with animals: if the poor could not be compassionate to their animals, then how could they be civil and humane in society. For many early Victorians who embraced humanitarian reform, violence was no longer an isolated act, it was a symptom of an internal state of mind of the perpetrators. Ill-treatment of animals revealed a lack of social feeling and sentiment, and a clear predisposition to surrender to base passions. At a deeper level, all these concerns expressed a precarious feeling about the nature of the new social order. In a civilised society, feeling and morality supported each other in governing human relations. Inhumanity to beasts was represented as something which belonged to less civilised regimes in the past, along with judicial torture, mutilation, and similar barbarities. Pity, compassion, and a reluctance to inflict pain, whether on men or beasts, became a signifier of the degree of civilisation. By showing feeling towards animals, an individual revealed their capacity for moral sensibility. However, humanitarianism was not without its contradictions. While reformers had led several successful campaigns to remove ill-treated animals from the streets, their critics frequently expressed ambivalence about whether animal cruelty reform had made the streets any safer. Moreover, the reformer's argument that the only effective way to control rabies was at its origin in cruelty required nothing less than the complete reform of human–animal relations across the whole of society. By eschewing technical solutions, like confinement and muzzling, that halted transmission and arguing for the radical, seemingly unobtainable solution, reformers allowed the very attitudes and actions they were trying to reform to continue.

This ambivalence towards dogs was evident when the London Home for Lost Dogs opened in Holloway in 1860. This later moved and became the famous Battersea Dogs' Home.[105] A correspondent to the *Field* worried that it would actually foster cruelty by providing somewhere for ill-treated dogs to be abandoned and would generally give 'encouragement to the increase of under-bred curs that infect the metropolis'.[106] An editorial in the *Times* celebrated the banning of dog carts and other reforms pushed through by the RSPCA and stated, highlighting the changed sensibility, that there were now 'thousands of eyes to notice, and thousands of tongues ready to denounce' animal cruelty.[107] It welcomed the role of police and courts in bringing perpetrators of cruelty, and was proud that Britain was ahead of other countries in such reforms. However, the writer thought the Dogs' Home had taken humanitarian feeling 'from the ridiculous to the sublime' and that its supporters had 'lost their sober senses'. It went on, 'Why not a "Home" for lost £5 notes dropped in the streets?' and 'Why should there not be a home for rats?' Perhaps humanitarianism had reached a tipping point: it was acceptable to prevent cruelty to animals, but seeming not to promote kindness through charity. The re-emergence of rabies in the 1860s produced new versions of these issues.

3
Rabies Resurgent: 'The Dog Plague', 1864–1879

Between the early 1860s to the late 1870s two broad changes occurred in public, veterinary and medical views of rabies and hydrophobia. The perceived threat of rabies moved into the home; the 'Dog Days' of summer and canine madness affecting whole towns was gradually replaced in the public consciousness with the idea that there was a constant danger from pet dogs as well as strays on the street. Second, the two diseases were increasingly considered separately: rabies as an animal disease that might eventually be eradicated, and hydrophobia as a uniquely untreatable human condition with an intriguing psychological dimension. Indeed, hydrophobia was the only 'phobia' until the labelling of the new phobias, like claustrophobia and agoraphobia, in the 1870s. For most of the 1860s, veterinary and medical opinion on the diseases remained fragmented, with no group of professionals codifying and validating knowledge systematically. However, from the mid-1870s a network of veterinarians and doctors was formed around the campaigning of George Fleming, the editor of the *Veterinary Journal*, who became the national authority on rabies.[1] By 1880, this network had strong links with the government's veterinary officials, with doctors promoting new approaches to controlling infectious contagious diseases, and with proponents of the new laboratory medicine. There was also greater professional consensus and public acceptance that both conditions almost always had their origin in inoculation and that only very rarely, if ever, arose spontaneously. These changes coincided with a rising incidence of hydrophobia deaths, with peaks in 1866, 1871, and 1877, when they reached their highest ever level with 79 notifications. Hydrophobia deaths and rabies incidents remained relatively low in London (see Graph 3.1), and the problems became associated with the northern industrial

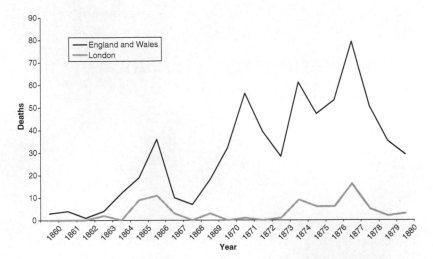

Graph 3.1   Hydrophobia deaths in England and Wales, and London, 1860–1880
Source: Compiled from *Annual Reports of the Registrar General's Office*.

towns of the West Riding of Yorkshire, the north Midlands and above all Lancashire. (See Map 3.1.)

'The mad dog crusade'

In 1862 only a single person died of hydrophobia in England and Wales; within four years the figure had risen to 36, with the majority of deaths in Lancashire where dog owning was growing in popularity among the industrial working class.[2] Rising wages allowed more dogs to be owned for sport, pleasure, companionship, or as a sign of respectability, while the swings of the economic cycle, for example, the cotton famine of mid-1860s, ensured that there were often large numbers of stray dogs on the streets.

On 4 June 1864 the *Liverpool Daily Post* carried a doggerel rhyme titled 'The Two Dog Shows', it began:

> All London for the last few days, as you, of course, well know,
> Has crowded Islington to see the Great Dog Show;
> Which has been totally eclipsed by Liverpool: we find
> They've had a dog show there for weeks of quite another kind,
> One which is 'open every day' in all the streets and lanes;
> And which consist of tortured dogs, and dogs without their
>   brains.[3]

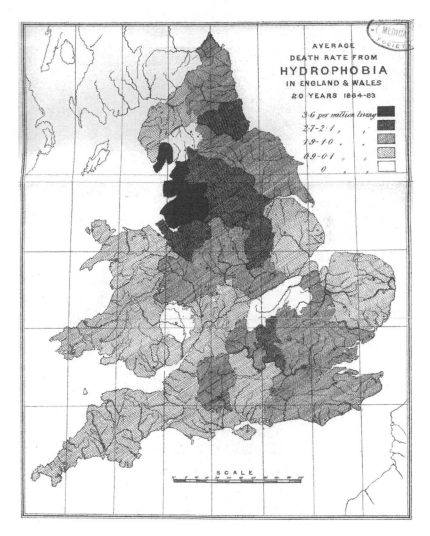

Map 3.1 Average death rate from hydrophobia in England and Wales by county, 1864–1883

Source: G. B. Longstaff, 'Hydrophobia Statistics', *Transactions of the Epidemiological Society of London*, (1885–1886), 5: 130–31.

Readers would have known that dog shows were a new and rapidly growing phenomenon, the first having been held in Newcastle in 1859, while mad dog chases and their gory endings were only too familiar. Many contrasts were implicit in the poem: between order and chaos,

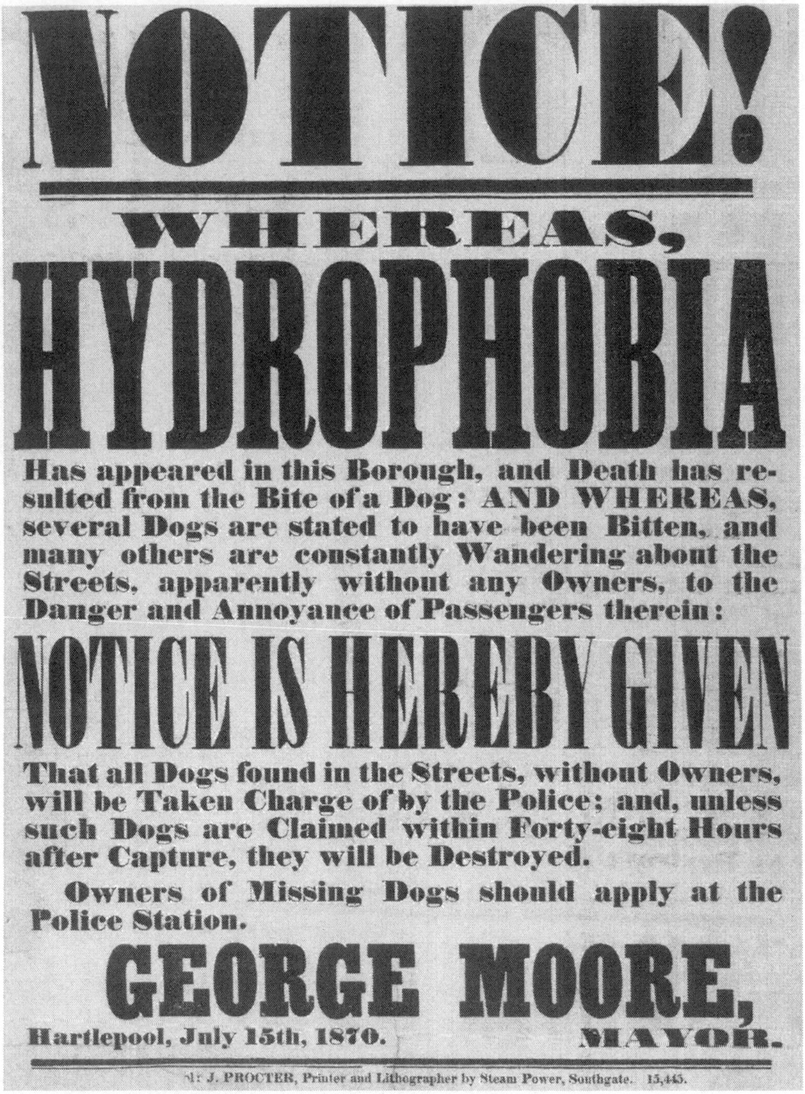
NOTICE!

WHEREAS,

HYDROPHOBIA

Has appeared in this Borough, and Death has resulted from the Bite of a Dog: AND WHEREAS, several Dogs are stated to have been Bitten, and many others are constantly Wandering about the Streets, apparently without any Owners, to the Danger and Annoyance of Passengers therein:

NOTICE IS HEREBY GIVEN

That all Dogs found in the Streets, without Owners, will be Taken Charge of by the Police; and, unless such Dogs are Claimed within Forty-eight Hours after Capture, they will be Destroyed.

Owners of Missing Dogs should apply at the Police Station.

GEORGE MOORE,

Hartlepool, July 15th, 1870. MAYOR.

J. PROCTER, Printer and Lithographer by Steam Power, Southgate. 15,445.

Figure 3.1 'Notice! Hydrophobia', 1870

Source: PortCities Hartlepool, Robert Wood Collection, ID: 2557, John Proctor, 15 July 1870. Courtesy of PortCities Hartlepool.

refinement and brutality, breeds and curs, respectable and disrespectable classes, and between North and South. Commentators in London and in the provinces struggled to reconcile events in Liverpool, and then across Lancashire, where dogs were slaughtered on a seemingly industrial scale, with the idea of the new humanised dog.

Reports of mad dog incidents and deaths from hydrophobia in Liverpool mounted in the spring of 1864. The town council responded by putting up posters and issuing control orders, and in a matter of days nearly a thousand dogs were reported to have had been killed by the police, and hundreds of dog owners found themselves in court facing fines. (See Figure 3.1.) Editorials in the *Liverpool Daily Post* resolutely condemned 'the mad dog crusade', maintaining that while there were genuine instances of the disease, there was nowhere the number of rabid dogs that had been reported.[4] Letters to the local press stirred the controversy with headlines such as 'The Insane War Against Sane Dogs' and 'A Plea for the Dogs'. Many concerned residents would have agreed with the view advanced by Philo-Canis, who, with a clear reference to hysteria, condemned the 'absurd panic' that has 'seized the females of both sexes'.[5] An editorial in the *Liverpool Daily Post* argued that popular fears and irrational views had flourished in the absence of clear advice from doctors and veterinarians.[6] An Alderman used the familiar comparison of the slaughter of innocent dogs to the burning of witches in the Middle Ages.[7] Moreover, it seemed that the very measures taken to reduce the threat had the reverse effect, over stimulating imaginations and fanning the flames of anxiety. One critic explained that prior to the 'crusade' women were generally not known to suffer from hydrophobia as they spent less time on the street; now they were as vulnerable as men.[8]

Some commentators worried about spreading moral contamination, as the respectable classes, along with women and children, witnessed bloody scenes in which the public seemed to lose all civility. One dog lover agonised over the loss of sentimentality in the streets when he wrote:

> The dog is property, an Englishman's or an English woman's ... He is the embodiment of the most perfect faithfulness: he never is treacherous, never betrays; his constancy is of the most devoted character, even when harshly treated and instantaneously sensitive to a kind word or kind act, however ill used he may have been before; the most forgiving, perhaps of all create things ... Take, then, my dog and

destroy him: you not only destroy my property which is an Englishman's but you painfully hurt my feelings.[9]

The degree of feeling for dogs in this quote made the matter-of-fact manner in which dogs were dispatched all the more shocking for many citizens. So too did the tendency of the Liverpool poor to treat the cry of 'mad dog' as an opportunity to form mobs.

I am really shocked at even dogs being killed in public streets in a civilized town like Liverpool. I am a foreigner, and have visited many countries, but I have never heard of such cruelty barbarous and scandalous proceedings against the poor canine race as I see and hear of in Liverpool.[10]

Such complaints were clearly indicators of the social and moral distance that many middle-class inhabitants felt from the Liverpool poor, yet they also revealed a disappointment that the seemingly growing respectability of the working class was just a thin veneer, beneath which still lurked elements of the dangerous classes of previous decades. These sentiments were evident in the closing lines of the *Liverpool Daily Post's* poem quoted above.

And yet we cannot now walk out but some poor cur limps by
With lolling tongue, and hanging tail, and howl of agony.
And if into a corner he at last contrives to slip.
He's speedily dislodged by 'half a brick' or by a whip.
A crowd of ruffians follow him, 'Mad Dog, mad dog' they call –
Well, if he is not after that, he ought to be that all!!!!
One sees a man upon the road engaged in labour hard,
While on his jacket lies dog, and over it mounts guard.
Then seeing him round and cracks his skull against a post or rail.
And why such scenes? Because some few of late have grown so
    very
Shakespearean that they models make of Verges and Dogberry.
'Up, Butterman, and at them', such brutality put down,
Which brings disgrace and ridicule each day upon the town.
This you can do by giving out that you will ne'er allow
That man to taste your butter who ill-treats a poer 'Bow-Wow.[11]

The terror, real or manufactured, altered the Liverpool streetscape and seemingly turned ordinary people into murderers.

While ruffians were said to be the source of the panics, it was often respectable owners who found themselves in court being mocked by their 'idle and disorderly' inferiors, who had allowed their dogs to be killed rather than admit ownership and risk being fined.[12] The middle class defended themselves by arguing that their social position insured that they were responsible citizens who always had their dogs under control, even though they were technically at fault for not using a lead or a muzzle.[13] One defendant claimed that their dog was a harmless pet, the size of a cat and entirely without teeth; another said that his dog's muzzle had been stolen.[14] Some working-class owners did appear in court and there were often humorous exchanges, not least when offenders of 'Irish persuasion' were in the dock.[15] The respectable classes also felt threatened by the powers given to the police, who were not 'the sort of men' who could distinguish a well-bred dog from a mongrel. It was feared that the township would become governed on a spoils system as policeman found amusement and pleasure in killing the dogs that belonged to their social betters. On hearing that the police had developed a machine for drowning dogs in ever greater numbers, an editorial in the *Liverpool Daily Courier* expressed fear that the police would become 'rabid' with cruelty.[16] One correspondent to the *Liverpool Daily News* pleaded for a Home, which unlike the one in London, would cater, not for shabby mongrels and curs, but the 'innocent specimens of the canine race' and the 'dog that may have simply escaped from the drawing room to the street, or that a valuable Newfoundland who may have been gladdened a mother's heart by rescuing her child from a watery grave.'[17]

While rabid dogs, barbarous mobs, and vindictive policemen were all viewed as dangerous, so were the material consequences of their actions. The headline, 'Another Mad dog More Dangerous Dead than Alive' drew attention to how one local authority had ignored the problem of dead dogs.[18] The writer went on, 'in a thickly populated neighborhood, with this putrid carcass, lying around for weeks uncovered during this unusually hot weather, will soon make it a question whether the mad dog is not more dangerous to the health of the neighbourhood when dead than if alive'. The sight of canine corpses was also distressing. One resident was shocked when he found that a policeman had thrown two dogs into his ash pit, one half dead and still kicking, which took an hour to die.[19] Putrid canine carcasses were also connected with spreading disease to other animals; for example, dogs drowned in water troughs were thought likely to poison cattle and horses.

There was little national press coverage of the Liverpool outbreak, but the event was eagerly reported by neighbouring towns. The large number of dogs killed in Liverpool led the *Manchester Guardian* to dub the bloody spectacle the 'Liverpool crusade against dogs', while the *Stockport Advertiser* called it the 'Liverpool mania'.[20] Control measures and their social repercussions quickly spread across Lancashire. The Mayor of Chester issued a notice requesting persons in the city to keep their dogs confined from May to September, hoping that it 'will prove a sufficient warning to the owners who have hitherto been accustomed to allow their pet animals to roam at large through the thoroughfares'.[21] Yet, by mid-June, Manchester and Stockport succumbed to 'panics', with the authorities issuing notices of confinement and muzzling orders. The intensity of the public reaction to rabies bemused many contemporary observers, especially as thousands of dogs were destroyed before the arrival of the 'Dog Days'.[22]

An indication of the pervasiveness of rabies at this time can be seen in the way it infected political debate. On 6 July 1865 – at the start the 'Dog Days' – Palmerston's liberal government resigned and William Gladstone stood down from his Oxford University seat to seek re-election in the South Lancashire constituency that centred on Liverpool. The leaders of his party were worried about the move as it would give him more scope to express his radical views; indeed, Palmerston was reported to have said, 'Mr. Gladstone is a dangerous man. In Oxford he is muzzled; but send him elsewhere, he will run wild.'[23] In response Gladstone told an election meeting in Liverpool, 'At last, my friends, I am come among you; and I am come – to use an expression which has become very famous, and is not likely to he forgotten – I am come unmuzzled.'[24]

'The Dog Plague'

The years 1865 and 1866 were years of general pestilence in Britain: the public had to endure the fourth visitation of cholera; farmers and live-stock suffered from the cattle plague and rabies reached levels last seen in the 1830s.[25] While anticontagionists maintained that there was some malign, possibly general epidemic influence over the whole country, most doctors and veterinarians saw all three plagues as communicable.[26] With cholera, there was greater acceptance of John Snow's water-carriage theory, with cattle plague there was no doubt it was highly contagious, and with rabies professional opinion had moved to accept contagion for almost every case.[27] Public views on all three diseases were more varied, with notions of communicability mixed with spontaneous generation and immaterial influences. Few links were drawn between the two great scourges of cholera and cattle plague, but dogs and rabies were often associated with the other diseases. This possibility was raised by 'Bow-Wow' in a letter to the *Stockport Advertiser* in May 1866. He explained that there was a plague mania in the country: 'We are just now surrounded with all sorts of plagues – plagues animal, plagues local, and plagues national'.[28] So many allegedly mad dogs had been destroyed that they had been allowed to become 'stinking carcases' and there was a danger 'of disseminating offensive gas'. The situation was so appalling that the same author reported that 'we shall thereby have the first instalment of the Asiatic cholera in this well-regulated and decidedly-healthy borough – all through the Dog Plague'.

Roaming curs were blamed for spreading disease amongst cattle. For example, William Law wrote to the *Times* that,

> the peril to our herds arising from the dogs of tramps, being allowed to roam at large by day, and to prowl everywhere unchecked, unseen by night, is, I am convinced, one the greatest which now threatens our rural districts ... we feel that we are at the mercy of these plague spreading curs ... They defeat every system of precaution, they break the cordon of all attempted isolation, they mar the wisest schemes, and establish a medium by which infection and healthy cattle are constantly brought *en rapport*.[29]

Many stray dogs were rounded up with other animals under the stringent regulations; for example, Cattle Plague Prevention Committees in Middlesex ordered the destruction of stray dogs.[30] Mad dogs were reported in the countryside as well as in the towns, as a threat to man and beast. A letter in the *Times* in May 1866 reported that 'about six weeks ago a cow of mine was bitten on the nose by a stray dog; yesterday, she became restless, and this morning was so fierce and violent that I sent for a veterinary surgeon who, like myself, was of opinion that she was suffering from hydrophobia'.[31] The image of the rabid cur spreading animal madness amongst cattle and sheep deepened the sense of the calamity and catastrophe at large in the country.

The centres of the 1865–1866 rabies outbreaks were Lancashire and London. In the affected regions, local authorities implemented control measures and there was continuing public division about the use of muzzling, seizure, and slaughtering. A letter in the *Times* from 'Dogmanity' in June 1865 showed that the measures against dogs had struck a new sensibility. The writer cited a 'letter' he had received, from a dog, urging the 'extinction of the lower classes of the canine race', which would end 'the protracted martyrdom of dying by inches from hunger, thirst, and horrible anxiety and agony of mind'.[32] Dogmanity, while agreeing that death was preferable to the muzzle, maintained that street curs were not without character, feelings, and intelligence. In Manchester the public mood over dogs switched between fear and sympathy. L. Danby wrote to a local paper of a population under siege,

> [T]he number of dogs known to have gone mad in this district, and that have been at large and spreading the contagion, is something fearful to contemplate. There are many families in the neighbour-hood of Brook's Bar, Whalley Range, and Stretford that dare scarcely leave their houses, and whom necessity compels with fear and trembling.[33]

Yet, there were also complaints that the police were killing dogs indis-criminately, showing no mercy, making no effort to find the owner and allow them to pay any fine.[34] Panics returned to Liverpool in 1866 with a novel aspect to the slaughter, the borough engineer reported that in a single week, as well as 297 dogs, 30 cats, 1 sheep and 1 goat had been collected from ashpits in the town.[35]

All three plagues prompted new legislation to try to halt future visit-ations: in the wake of cholera – the Sanitary Act, 1866 was passed; the cattle plague led to the passage of the Contagious Diseases (Animals) Act (CD(A)A), 1866; and with rabies a Protection Against Dogs Bill was laid before Parliament. This Bill was introduced by Sir Colman O'Loughlin, MP for Clare, and was aimed largely at the problem in Ireland; it con-tained measures to make dog owners liable for injuries to people and property, especially farm animals. It also gave powers to the police to round-up and kill unwanted stray dogs. While the legislation was not specifically aimed at rabies, its supporters included specific anti-rabies measures because of the precedents in similar legislation, the recent experience of the disease, continuing public fears, and in the hope of garnering political support. Rabies lurked in the background of almost all legislation concerning public order and regulation of the streets. Supporters insisted that there was little deterrent to owning vicious dogs and that ignorance led the public to be unconcerned about the threat of mad dogs. The Bill passed the House of Commons easily, but proved controversial in the House of Lords where opposition was led by Lord Cranworth, on the grounds that there was not the slightest necessity for the legislation, as there were already three Acts of Parliament in force for the protection of sheep from dogs, and there were penalties under cur-rent law to fine anyone who permitted a savage dog to be at large.[36] In the following year, O'Loughlin decided not to re-introduce the Bill; his cause would not have been helped by the fact that the number of deaths from hydrophobia fell sharply from 36 in 1866 to 10 in 1867.

Thwarted with specific legislation, those concerned about the welfare of the public and their dogs returned their attention to the Dog Tax.[37] Their claim was that the Dog Tax was only paid by responsible, law abiding owners, exactly the people whose dogs were least likely to be cruelly treated and to spread rabies. Reformers argued that the tax was too easily evaded; estimates were that only 300,000 out of 3 million dogs were covered.[38] They added their voice to longstanding criticisms that the tax was too complex, with too many exemptions based on the economic value of the animal and its breed. According to one parliamentarian, if a Dog Tax collector 'is to take the statement of the dog owner, or that of his neighbour, he will conclude that the dog is exempted'.[39] Proposals to banish exemptions were justified because reckless and irresponsible members of the lower ranks could easily evade the tax by telling bare-faced lies.

In response to calls for change, in April 1867 the government abandoned the Dog Tax and replaced it with a flat rate licence. They also reduced the duty payable from the 12s to 5s. The hope was that more dog owners would take up a cheaper, simpler, and fairer system, and that the streets would be rid of 'many of the useless portion of canine species'.[40] However, there were new criticisms. Some opponents feared that at its new lower level it would not cover the costs of collection and the government would abandon it, others worried it would in fact enlarge the stray dog population as poor dog owners would struggle to afford the fee. A self-proclaimed 'Sufferer' reported that, 'as any one acquainted with the habits of a large class of the labouring population of this neighbourhood can testify, who may see daily wives and children ill-fed and worse-clothed, to enable their noble lord to feed his favourite watch dog on the best beef, mutton and bread'.[41] Other critics feared that, unable to afford the licence, many poor dog owners would let their animals loose; indeed, there were soon reports of streets infested with large numbers of strays.[42] A report in the *Lancet* pointed to specific consequences of the new duty, it 'produced a terrific slaughter among the members of the canine race, and so many are thrown into the Thames that the effluvium is very great'.[43] The Thames police appeared to ignore the hazard, and by August one angry commentator felt compelled to call attention to a new pestilence, 'the disgusting stench of an animal's carcass drifting up and down the river under a scorching sun'.[44]

'Mayne's Law' and the Dogs Act, 1871

Animal welfare reformers welcomed the inclusion of clauses in the Metropolitan Street Act, 1867 that allowed the police to round up strays and impose muzzling orders when rabies was suspected. The legislation followed a Select Committee on the traffic problems of London, and the Chief Commissioner of Police's request for powers to control the stray dogs that were notorious for annoying draught horses and sometimes causing them to bolt, leading to traffic accidents and casualties.[45] There was no consultation with veterinarians or doctors, the issue was principally one of public protection and safety; indeed, the new law extended police powers to act against any stray, not just those 'reasonably supposed to be mad'.[46] Chief Commissioner Richard Mayne first used the new powers – eventually known as Mayne's Law – in the summer of 1868, seemingly pre-emptively. There were few reports of rabies and hydrophobia in London, so his motives were about the social order rather than public health. He introduced muzzling and the round up of strays, using a special constabulary force armed with lassos and gauntlets.[47] A *Daily News* correspondent commented on the immediate effects of the order: 'an unmuzzled dog is becoming a far great rarity than a black swan and the West End and suburbs of London have fewer disorderly curs than at any time within living memory'.[48]

Public reactions to the measures were divided. Middle-class dog owners welcomed the removal of strays and the dogs of the poor, but feared that the police were 'stealing' well-bred dogs and were convinced of the invulnerability of their animals to contagion. Many disliked the obtrusive nature of the muzzle and suggested that muzzling spread rabies amongst their cherished pets by causing frustration and anger, and allowing saliva to be reabsorbed.[49] As ever, there were allegations of a 'dog mania' and of the police were taking innocent dogs. Newspaper editorials guessed that few of the dogs killed on the streets were actually rabid. The *Standard* referred to the 'Dog Slaughter' and the *Daily Telegraph* asked the Royal Society for the Protection of Animals (RSPCA) to take immediate action against the police, as too often, 'A yelp of anguish, a howl of despair, a moan of entreaty is heard in the street ... ; four-footed friends are torn ruthlessly away if not accompanying their masters or mistresses'.[50] The medical press was not necessarily more supportive of 'Mayne's Law'; an editorial in the *Lancet* entitled 'Legalised Cruelty' condemned muzzling and insisted that it did little to prevent the spread of disease.[51] Nor were veterinarians necessarily on board. James Allen, a Government Veterinary Inspector, admitted that it was very rare in veterinary practice to come across a real case of rabies, and that a very great number of dogs were 'condemned by incorrect diagnoses'.[52] He also repeated the familiar complaint that the reported numbers of mad dogs were exaggerated by the 'very dread of hydrophobia'.

So overwhelmed were the police with the task of holding, killing and disposing of dogs, and by the public reaction to the early weeks of 'Mayne's Dog Law', that they turned to the Home for Lost and Stray Dogs for help.[53] The Home became the capital's depot for stray dogs, receiving hundreds of animals. The press were quick to recognise the symbolic importance of the Home in providing hope that owners would be reunited with pets that had been lost during the dog mania or 'wrongly' picked up by the police. The press portrayal of the Home was often equivocal; a correspondent in the *Morning Post* reported at length on what he termed this 'great slaughterhouse', but tried to reassure respectable owners that 'If the animal is well bred, it will be spared' and marvelled at the ability of the head keeper, Pavitt, to wean out refined dogs from mongrels.[54] Owners were given three days notice to claim their dogs, but in less busy times this was period was extended. If not claimed, well-bred dogs were put up for auction; mongrels were not so lucky, they were destroyed.

Mayne's Law seemed to keep rabies at bay from London, a situation shown by the number of deaths from hydrophobia in the capital: 0 in 1868, 3 in 1869, 0 in 1870, and 1 in 1871.[55] However, in the rest of the country deaths were rising, from 7 in 1868 to 55 in 1871. The main foci, of what veterinarians saw as a new epizootic, were again Lancashire and the West Riding of Yorkshire. One indicator of the change was the reappearance of harrowing accounts of terrible attacks and agonising deaths in the press. In November 1868 in Halifax, five deaths were attributed to hydrophobia over a three months period.[56] On 2 December 1868, the *Preston Guardian* reported that five people had been bitten in the town, and three months later one, a 13-year-old boy, died.[57] Things remained tense in the town and fears escalated when in March 1869, a congregation worshipping at Fulwood Church heard a noise at the door and turned round to see a large rabid dog at the entrance.[58] Later that month, 15 mad dogs were reported to be loose around the town. In both places the local authority introduced measures against dogs, a pattern repeated across northern England.

In mid-1870s hydrophobia deaths returned to the south-east.[59] The rising tide of reports led reformers to try once again to bring greater uniformity, coverage, and strength to police measures. This time they were pushing at an open door as the Dogs Act 1871 was passed by Parliament with little debate. The Act extended police powers nationally along the lines of the Metropolitan Street Act, 1867, allowing the seizure of 'dangerous dogs' that were not 'under control' and giving local authorities the powers to introduce controls. However, implementation was left to the discretion of local officials, which kept open the possibility of rabid dogs entering 'controlled' towns on their 'march' from adjacent, 'uncontrolled' districts.

Soon after the Act came into force, spokespersons for rural interests began to complain that it was designed for an urban problem, but was now being used inappropriately in the countryside. The opposition of elite sportsmen was aired in the pages of *The Field*. One correspondent saw the legislation as an encroachment on his liberties and property, and argued that protection of their hounds had primacy over any right of public protection against stray dogs. Sportsmen feared that the predilection of hunting hounds to fan out when pursuing their prey could lead to their dogs being wrongly identified as strays.[60] There were also class anxieties; another correspondent 'feared that the new legal framework left gentlemen vulnerable to the grudges of the criminal poor or to the disgruntled poachers that held a grudge against being caught by making the allegation that the dog that had bitten him was mad'.[61] The attack on privilege was symptomatic of a political crisis since the widening of the franchise in 1867 and a deep distrust of how those interests were being represented in the House of Commons.

The leading voice of rural protest was George Jesse, from Henbury, near Macclesfield in Cheshire, who was a popular writer on the dog.[62] In 1871, he set up the Association for the Protection of Dogs and Prevention of Hydrophobia, the main aims of which were to repeal the dog muzzling law; to publish information about rabies and hydrophobia; and 'to render thereby them of still rarer occurrence'.[63] Jesse repeated the familiar claim that rabies was hugely over-reported and in fact was exceedingly rare. Indeed, some went so far as to suggest that it was purely imaginary and that all cases were misdiagnosed instances of tetanus or epilepsy. Jesse appealed to the friends and owners of dogs, to masters of hounds, to sportsmen, and others to join his Association. He complained that a muzzled dog could no longer defend himself from the 'rabble, or his master's person or property from the thief'.[64] What was needed to control rabies was legislation that rooted out cruelty to dogs, though his real venom was reserved for the dog stealer, who chained up dogs and fed them such unwholesome food so that they became rabid.

Another critic, 'Beth Gelert' – a pseudonym taken from the heroic dog of William Stewart's poem of that name published in 1811 and may have been Jesse himself – maintained that the Dogs' Act was 'a sad proof of the degeneracy of Englishmen' and symptomatic of a wider social and cultural malaise.[65] This letter was first published in the *Macclesfield Guardian* in 1871, but gained wider recognition when it was included in Ruskin's letters to the working men of England, *Fors Clavigera* in April 1874.[66] The author complained that the Dogs Act showed how democratic Britain had become governed by a ruling class that was exploiting the working class and ruining the country. The Act signalled a dangerous, callous government, 'not even seen in the times of Canute or Edward the Third's ruthless reign of England'. 'Beth Gelert' wrote, 'why all this hubhub, this epidemic terror, about a disease which causes less loss of life than almost any other complaint known, and whose fatal affects can, in almost every case, be surely and certainly prevented by a Surgeon?' 'Beth Gelert' was bemused by the legislation against a disease with such low mortality, when compared with 'boiler and colliery explosions, railway smashes, and rotten ships, to the overcrowding and misery of the poor; to the adulteration of food and medicines; to the sale of fermented liquors. Also to dirt, municipal stupidity, and neglect, by which one city alone – Manchester – loses annually above three thousand lives'.[67]

The new profile of rabies and hydrophobia found expression in Richard Doddridge Blackmore's *Maid of Sker* published in 1872. Blackmore is best known for his novel *Lorna Doone*, but he regarded *The Maid of Sker* as his best work. In the denouement the villain – the malevolent Parson Chowne – was killed off by no ordinary hydrophobic convulsion. Blackmore uses dogs and rabies throughout the novel to aid the depiction of eighteenth century Devon as a 'gloomy and devilish place', and to define the viciousness of Chowne's character.[68] At the climax, the narrator, Davy Llewellyn goes to the Parson's home and finds him 'barking, howling, snapping of teeth, baying as of a human blood hound, froth spluttering of fury, and the smothered yelling'.[69] The Parson's latest wife – that he had many also told of his character – asked if she could help control the fits, but when Davy went into the room he was confronted by a shocking sight:

> It was Chowne's own dining-room, all in the dark, except where a lamp had been bought in by a trembling footman, who ran away, knowing that he had brought this light for his master to be strangled by. And in the corner lay this master, smothered under a feather-bed; yet with his vicious head fetched out in the last rabid struggle to bite. There was the black fair, black face and black tongue, shown by the

frothy wainscot, or between it and the ticking. On the featherbed lay exhausted, and with his mightily frame convulsed, so that a child might master him, Parson Jack Rambone, the strongest man, whose strength like all other powers had laid a horrible duty upon him. Sobbing with all his great heart he lay, yet afraid to take his weight off, and sweating at every pore with labour, peril of life, and agony.[70]

In smothering his fellow Parson, Jack Rambone was resorting to a practice that Victorians associated with earlier, less civilised times, but which still worried the popular imagination. However, in this episode Blackmore gives rabies and smothering moral meanings. Davy remarked that 'The biggest villain I ever knew showed his wit by dying of a disease which gave him the power to snap out at the very devil'.[71] In the process of smothering Chowne, Rambone was bitten several times, but felt the experience was 'a stroke towards his own salvation' and that in performing the 'horrible job he earned repentance, fear and conscience'.[72] Blackmore's use of hydrophobia confirms that, while attempts to control rabies and its human consequences were becoming increasingly concentrated in the legislation and in the courts, the disease retained wider cultural meanings and resonances.

George Fleming and epizootic rabies

Deaths from hydrophobia kept rising, from 7 in 1868 to 56 in 1871, and then fell before reaching 61 in 1874. These deaths, with the many more mad dog incidents that accompanied them, gave both diseases a high public profile. Deaths continued to be concentrated in Lancashire, the West Riding of Yorkshire, Cheshire, Staffordshire, and London. As previously, there were many melodramatic narratives of dogs running amok and press reports of the agonies of hydrophobia. For example, in 1874 the Manchester City Coroner heard a case of night watchmen who was so violent that he was placed in the lunatic ward and then a padded cell.[73] In Newton Heath, near Manchester, the police were sent for after a poor, naked man suffering from hydrophobia drove his family out of their house and forbade anyone to enter. He attempted to strike a policeman with a poker and eventually flung a board through a window.[74] Such was the problem in Manchester that a local businessman, Mr Reilly, the proprietor of the Pomona Gardens pleasure grounds, put up £500 to encourage a search for a cure.[75] In Macclesfield, Cheshire, the police were armed with pistols and bludgeons in preparation for any incident following the destruction of six mad dogs in six weeks.[76] In London, the cry of 'mad dog' in Newington Causeway led to a pandemonium in the thoroughfare as a wolfhound ran its 'mad career', snapping at passers-by, 17 of whom had their wounds cauterised in local chemists' shops.[77]

In 1874 controversy developed between the up and coming veterinarian, George Fleming and the aristocratic sportsman Grantley F. Berkeley on the nature of rabies in dogs and what to do about the rising incidence. Fleming was a veterinary officer in the Royal Engineers, who developed an interest in rabies while writing his encyclopaedic two-volume *Sanitary Science* eventually published in 1875.[78] (Figure 3.2.) Again, like many of his peers, his views of animal diseases had been transformed by the cattle plague and its aftermath, and he became devoted to the cause

GEORGE FLEMING, C.B., LL.D., F.R.C.V.S.,
LATE PRINCIPAL VETERINARY SURGEON OF THE BRITISH ARMY.

Figure 3.2   George Fleming, 1901

Source: *Veterinary Journal*, 1901, 3: facing page 307. Reproduced by permission of the Wellcome Library, London.

of suppressing contagious animal diseases through sanitary and legislative measures.[79] Grantley F. Berkeley represented, often in caricature, the rural interests that were increasingly worried about the implications of a contagion model of rabies for their sports. He was an eccentric aristocratic sportsman who set out to expose 'veterinary error' on the question of rabies.[80] He characterised the views of Fleming and other veterinarians as a 'fiction', arguing that distemper was often mistaken for rabies, and that its origins lay in ill-treatment. He claimed that 'the fancy' and sportsmen like himself knew more about dog diseases than veterinarians and doctors, and that they had not been fooled into mistaking rabies for other common diseases.[81]

Fleming first published his volume entitled *Rabies and Hydrophobia* in 1872. It covered everything from their history, through their geographical spread, causes, symptoms, pathological anatomy, their 'contagious properties', and how to suppress rabies and prevent hydrophobia. He argued for rabies to be treated as an epizootic – an imported animal plague – and be added to the CD(A)A. In other words, he wanted it to be stamped out on the model of the cattle plague in the 1860s. He also disseminated his views through the *Veterinary Journal* which he founded in 1875 and edited. Although Fleming followed in the tradition of Blaine and Youatt, he wrote as though he was starting anew and needed to tell his readers everything there was to know about rabies. Above all he aimed to fill the gaps and remove the uncertainties created by there being no authoritative cadre of professionals working on the diseases. Veterinarians continued to show little interest in the dog and its diseases, and as we have seen, the vacuum was filled by the new breed of authors writing for dog fanciers, breeders, and devoted owners. There was also relatively little interest amongst doctors; hydrophobia was very rare, most practitioners never saw a case in their career, and although classified as a zymotic disease, it was never on the agenda of the public health movement. Thus, the principal repository of knowledge on rabies and hydrophobia was the popular memory and imagination. Every town and village chemist still had a recipe for dog bite victims, and practices such as eating the biting dog's heart or sea-dipping continued to have currency. Fleming, medical men, and local authorities repeatedly bemoaned public ignorance and superstition over rabies, but the fears and anxieties over horrid deaths seemed well founded to the man in the street. Public perceptions of risk are never 'rational' or purely calculative, but were, and continue to be, based on a whole range of factors.

Fleming and medical men who wrote on the matter, such as John Burdon Sanderson, the Director of the Brown Animal Sanatory Institute in London, who was also a leading physiologist and medical officer of health, were concerned that too often the early symptoms went unnoticed and put people in dangerous situations.[82] One goal was to educate the public, which was evident in Fleming's repeated calls to have the symptoms of rabies printed on the dog licence. He divided the development of rabies into three stages that could quickly pass from one to another without an owner noticing. He warned that contrary to popular opinion it did not begin with signs of raging madness, fury, and destruction; rather, the dog sought isolation and retired from company, though such behaviour could be accompanied by extraordinary bouts of affection, even sexual excitement.[83] Thus, the key thing for owners to watch out for was a change in the dog's demeanour.[84] In the next stage these same symptoms become more marked, though the animal also began to hallucinate and snap at imaginary foes. Only in the final stage did the dog become aggressive, and then only with other animals, or if confronted.

All dogs did not develop 'furious rabies'; there was also paralytic or 'dumb rabies', where the animal became morose and passive. (Figures 3.3 and 3.4.) The latter condition, at least to the public, was new, further complicating the disease in the public mind. Thus, rabies and the dogs it affected had become chameleon-like, and even more threatening. Fleming maintained that at every stage the dog's 'instinct impels it, at times, to draw near to its master, as if to ask for relief from its sufferings'.[85] Indeed, so powerful was domestication that the voice of the owner could subdue a rabid dog even in the final, furious stage, and that sometimes a dog would go on its 'march' round the streets or across country, rather than inflict rabies upon its 'family'.[86] When explaining its unusually high incidence in contemporary Europe, especially in France and Germany, and more recently in Britain, Fleming blamed domestication, contrasting its low incidence in the East where dogs were undomesticated and lived naturally in the wild.[87] He complained about the use of the term 'hydrophobia', which he wanted to censor because it wrongly suggested that the primary symptom of rabies was the dread of water, when it was hardly ever found in animals.[88]

When discussing how rabies developed and spread, Fleming maintained that 'in 999 out of 1000 cases it was spread by inoculation'.[89] This statement would haunt him in years to come, as his opponents on rabies policy repeatedly used this tacit admission of a 1 in 1000 chance of spontaneous development against policies of stamping out.[90] Working within the framework of direct contagion and the new germ theories of

Furious Rabies : Late Stage.

Figure 3.3 'Furious Rabies: Late Stage'

Source: G. Fleming, *Rabies and Hydrophobia*, 1872, 230. Reproduced by permission of the Wellcome Library, London.

" Dumb Madness." Drawn from Life. (Sanson.)

Figure 3.4 'Dumb Madness'

Source: G. Fleming, *Rabies and Hydrophobia*, 1872, 232. Reproduced by permission of the Wellcome Library, London.

disease, he acknowledged that 'nothing has been discovered in the morbid salvia or foam to account for its poisonous properties, the micro-scopist searches in vain for the secret of its death-bringing power'.[91] This situation was not unusual in the early 1870s; after all the 'virus' of one of the most contagious and common such disease – smallpox – had not been isolated or identified. Like most of his peers in veterinary medicine and medicine, Fleming's belief in contagion came from experience in the field and in the clinic.[92] Nonetheless, rabies was different to other contagious diseases in not being as specific as, say, cattle plague or small-pox: its incubation period was variable, its symptoms capricious, and its morbid pathology obscure. Furthermore, its sporadic incidence was still cited by remaining anticontagionists as proof that it could and did arise spontaneously. Fleming's answer to what happened in months and years between the outbreaks was to argue that its long latency period allowed it to 'carried' for months and even years in animals with a propensity to wander.[93]

In the 1870s Fleming was associated with three measures to control rabies.[94] First and foremost, he recommended the culling of 'useless curs', as these were the main source and spread of rabies. He agreed with popu-lar views on the menace of the 'currish brutes of the poor who allow their dogs to be a public nuisance'.[95] Second, he argued that liable dogs should not only be taxed more rigorously, but that collars should bear the mark of payment and the identity of the owner; he even suggested that a dog census might be created. This implied a high degree of public surveillance and bureaucracy, which would require the state's intrusion into homes. Third, and as noted already, he recommended that the symptoms of rabies should be listed on the back of all dog licenses. His views on muz-zling were ambivalent. He preferred other methods, but felt that muzzling had to be used when rabies was prevalent in an area.[96] Even then, the greatest gain with muzzling was that it served as a badge for the well-cared-for dog, whereas dangerous dogs were invariably without a muzzle, being either ownerless or having been left to their own devious devices on the street. As Youatt before him, Fleming asked the public not to kill the suspected dog, recommending its isolation to discover, first, whether the animal actually had rabies and, second, to prevent unnecessary anguish in those unfortunate enough to have been bitten.

Fleming came under attack from Grantley F. Berkeley as typical of the new scientific and secular experts who were challenging traditional sources of knowledge and power.[97] He penned a number of letters to the *Times* and the *Field* on rabies, and elaborated further in his book *Fact Against Fiction* published in 1874.[98] Press and veterinary reports showed that the 1870s rabies epizootic had spread to foxhounds, and regarded this as dangerous because these animals chased their prey across the countryside and were, by breeding, seeking to draw blood by biting. Foxhounds and hunting were, of course, symbols of the values of the gentry and aristocracy, and their traditional dominance over the countryside. If foxhounds had a propensity for spreading rabies, then the implication was that traditional elites were now a danger to the countryside and its people. Berkeley claimed that foxhounds rarely if ever suffered from rabies, contending that it only arose when dogs were maltreated, underfed, or chained up – not features of well-organised kennels. Indeed, he suggested that alleged outbreaks in kennel hounds were actually of distemper, a disease that was primarily miasmatic and affected younger dogs.[99] Having dealt with rabies, he turned to hydrophobia and suggested that the current public alarm was of nervous origin, affecting those with 'nervous temperaments', especially in the 'fairer sex'.[100] Showing typical condescension towards the public, he insisted that the growth in public apprehension about dogs followed from giving a 'little education' to the lower classes, a reference to the 1870 Education Act.

When the incidence of rabies reached a new high in 1874, influential figures from medicine came forward to back the calls for stronger controls. Many of these emerged from the cadre of public health doctors and scientific researchers who had been enrolled by Sir John Simon in the work of the Medical Department of the Local Government Board.[101] One of the most influential, John Burdon Sanderson, wrote to the *Times* in 1874 advocating the adoption of Fleming's three-part plan.[102] An editorial in the *Lancet* backed similar measures, though one supportive correspondent anticipated objections from groups opposed to state intervention, saying that he expected the emergence of a 'new crochet clique to agitate for the "question" of free-trade in hydrophobia and the "vested rights" of our canine friends to propagate it at their pleasure'.[103] The measures taken by local authorities varied, with reports of stringent measures in certain northern towns contrasted with the laxity of affairs in London.[104] At the start of the year the *Times* carried a pessimistic leading article, which contrasted the improved knowledge of most zymotic diseases with the position on rabies.

With regard to canine madness, we can scarcely be said to have any certain knowledge at all. We are absolutely ignorant of its essential nature, of its seat in the organism, of its causes when it occurs apart from contagion, of its modes of diffusion except by direct inoculation, of the circumstance conducive to its prevalence, or of the morbid changes to which remedies should be addressed.[105]

George Fleming must have been surprised and disappointed to read this assessment, but it again pointed to the situation where human zymotic diseases were being investigated by a growing network of epidemiologists, public health doctors, and experimentalists, while there were no equivalent networks for animal diseases.[106] And while Fleming's book was regarded as authoritative amongst the professions, his views continued to be attacked by those who spoke from real world experience rather than science. Their claim, that rabies was nowhere near as prevalent as police reports indicated, was repeated with the implication was that these groups had a vested interest in talking up its incidence.

The year 1877 saw the highest number of certified hydrophobia deaths in any single year in the nineteenth century. The incidence had risen in 1876 and in his Annual Report for that year William Farr suggested that this was due not to the increase in the number of dogs or neglect of police regulations but to a variation in the strength of its zymotic animal poison – lysine.[107] There was a reported panic in Glasgow in November after two men died, which led to the slaughter of 1,200 dogs.[108] In the year overall in England and Wales there were 53 deaths, 45 males and 8 females, a gender ratio that Farr attributed to women's usual workplace in the home or factory leading to less exposure to dogs.[109] In the peak year of 1877 there were 79 deaths, 61 male and 18 female, which were distributed around the country.[110] The largest numbers were in London (16) and the North-West (19) – mostly Liverpool and West Lancashire – with the remainder across southern counties; the West Riding of Yorkshire, where rabies was usually prevalent, only had a single death.[111] As early as January, the *Lancet* was warning that 'A panic may occur or be created without the justification of a peril, and it may produce the evil it madly dreads.'[112] And at the end of the summer an editorial pursued the same theme, the problem of distinguishing true hydrophobia from 'mental' hydrophobia – a functional cerebral disorder. Nonetheless, the journal called upon the government to act, backing Fleming's calls for greater public education and switching the control of rabies from the police and the Home Office to the Veterinary Department.[113] In October the *Times* carried an editorial reflecting on the panic of recent months and bemoaned the fact that the paper again had 'to try and discriminate between the truth of the matter and incrustation of fiction which has grown up around it'.[114] In early November on one day there were news stories on mad dogs in Devon, Kent, Warwick, Oxford, and Battersea.[115] The remedy suggested by the paper was not muzzles and chains, but to use taxes and other measures to restrict the ownership of dogs, to ensure that dog owners obeyed the law, and to make all dogs wear an identity tag on a collar.[116] (Figure 3.5.) An article in the RSPCA journal *Animal World*, entitled 'The Hydrophobia–Phobia', carried a warning about the 'multiplying forms of delusion' on the streets and stated that a major issue was to manage the public mind as the 'panic and unreasoning terror is not only needless, but in a high degree mischievous, and ought to be judiciously repressed'.[117]

The claim that rabies was epizootic across the country was challenged in 1877 by Thomas Scorobi, the manager of the Home for Lost and Starving Dogs. He reported that there had been no cases of rabies in strays collected since in 1870, the year from which the Home received all stray dogs collected by the police.[118] He stressed just how many times he and his colleagues had been bitten and that none had developed hydrophobia, though elsewhere he detailed the disinfection treatment

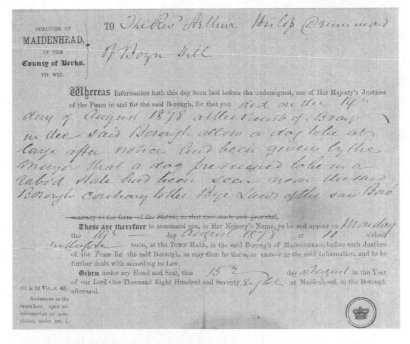

Figure 3.5 Summons issued by the Justice of the Peace for Maidenhead, Berkshire (under 11 & 12 Vict. c. 43) against Rev. A. H. Drummond (1843–1925) accused of allowing a dog to be at large in contravention of rabies by-laws. 15 August 1878. Reproduced by permission of the Wellcome Library, London.

he used for all dog bites. The Annual Reports explained how the Home isolated many dogs suspected of rabies, but almost everyone had recovered.[119] There was an exception in 1878, when 4780 dogs were brought in and one died of rabies, and that a few hours after its arrival.[120] Scoborio blamed public hysteria for the high profile of rabies, claiming that the people were easily alarmed when they saw a convulsing dog in the street.

> [T]he poor brute is consequently driven up one street and down another at his utmost speed, kicked, stoned, terrorised and maddened into fury, until he bites someone obstructing or pursuing him, whereupon without further evidence he is pronounced to be mad. Scenes like these are taking place in London perhaps hourly during the current panic.[121]

An editorial in the *Pall Mall Gazette* declared a 'mania for hydrophobia' was running wild through the country, stating that:

> Most of the deaths recently declared by frightened jurymen to be caused by 'hydrophobia' were, it seems to us, really caused by superstitious terror; 'died of fright' would have been a more appropriate verdict. An epidemic of nervous panic has been set up by hearing or reading sensational newspaper reports. If one case is reported in today's paper two or three more are sure to follow; and so it goes on, and may do until the panic has worn itself out or the newspapers have something else to write out about.[122]

According to the RSPCA journal *Animal World*, nearly all reports were the 'offspring of excited imagination' and 'are accepted as genuine only because at the present time the common-sense of our country is under the domination of terror'.[123]

Fleming was also criticised for denying 'the fact' that rabies was generated *de novo* and not just by his rural and animal welfare critics. The mid-1870s saw a debate in British science and medicine over the spontaneous generation of life, in which the *de novo* production of disease germs was a central issue.[124] It was in this context that Fleming's 1 in 1,000 admission was exploited. W. Lauder Lindsay, a physician at the Murray Institution for the Insane in Perth, argued against Fleming's 'mischievous errors' in denying that human actions were responsible for producing spontaneous rabies, through over-exciting their dog, from indulging it, and from causing it pain.[125]

Medicine and the epidemic of hydrophobia

There were important issues at stake for medicine and science, and opposition to spontaneous generation became a symbol of modern, progressive, scientific medicine. Such was the case in Joseph Lister's work to establish antiseptic surgery, and the efforts of medical officers of health to control of contagious and infectious diseases.[126] It was ironic, therefore, that it was it 85-year-old Sir Thomas Watson who emerged as the champion of the new possibilities of controlling contagious and infectious diseases.[127] In May 1877 he published an article in the popular monthly the *Nineteenth Century* entitled 'The Abolition of Zymotic Diseases', where he followed William Budd, John Snow, John Tyndall, and others in accepting that all cases of zymotic disease came from previous cases and that in principle all were preventable and possible to eradicate.[128] The review of his article in the *Lancet* felt that he had gone too far in denying the spontaneous origins of disease, showing the continuing power of this assumption in medicine.[129] Watson argued that it was now evident from epidemiological and pathological work that zymotic diseases could be effectively controlled by notification, isolation, disinfection and quarantines, though he warned that implementing such measures would involve struggles against the opponents of state intervention and the defenders of the liberty of a 'free-trade in disease'. In November 1877, Watson published a follow-up article in the same journal on 'Hydrophobia and Rabies', where he maintained that hydrophobia only developed from direct inoculation and could be eradicated by preventing the transmission of its germs or virus by adopting the plans of Youatt and Bardsley.[130] The denial of spontaneous origins was the basis of his analysis and recommendations, which included strict enforcement of the Dog Tax, the establishment of a universal quarantine for dogs, and a total prohibition of the importation during quarantine periods.

Watson's intervention signalled the beginning of a medical takeover of rabies and hydrophobia. This was stimulated by the continuing prevalence of the epizootic, the new prospect of eradication, and the possibilities of laboratory investigations. Fleming's 1872 book was republished in 1878, but he too was increasingly enrolled in medical groups and discussions. The new medical interest was evident in the creation of two committees to investigate the disease, one provincial and the other metropolitan. The first Committee was headed by Thomas Dolan, a general practitioner and Poor Law Medical Officer from Halifax and sponsored by the *Medical Press and Circular*.[131] Its report was published in 1878 entitled *The Nature and Treatment of Rabies or Hydrophobia*, with George Fleming writing the chapters on rabies and Dolan those on hydrophobia.[132] The report insisted, yet again, that the diseases had remained mysterious for too long and that this had intensified popular ignorance and superstition.[133] In this context, it was significant that Dolan still had to insist that the diseases were not imaginary, though he offered the 'modicum of consolation that the portion of real cases of rabies are comparatively few'.[134] He recognised that doctors had often confused cases of hydrophobia with mania, tetanus, and epilepsy, but again attempted to reassure any lay readers by stating that 'modern medicine has lifted mania, epilepsy, phthisis, and other diseases, from the sloughs of empiricism into the clearer regions of sciences. Rabies must follow'.[135]

Written by a clinician and targeted mainly at medical practitioners, Dolan unsurprisingly dwelt on the issues of diagnosis and treatment of hydrophobia. He wrote that that whilst there was no disagreement over the signs of its final stages, there remained uncertainty about early stage symptoms and whether a case was 'hysterical' rather than 'real'. Hence, it was not uncommon for doctors to make enquiries about the patient's personality and to consult friends and family to understand the circumstances of the alleged dog bite. If the dog was alive and healthy, the patient could be said clearly to be suffering from spurious hydrophobia, where doctors assumed that they had become obsessed with the bite and such patients could not be persuaded that a terrible and inevitable death was not imminent. Many doctors felt that such situations were fostered by media sensationalism, with the minds of the uneducated and those who lacked moral sense especially vulnerable. Spurious hydrophobia was sometimes portrayed as a form of mania, with children vulnerable, as they were to religious hysteria.[136] With men, wives were often blamed for transferring their innate anxiousness about possible death to their husband.[137] Sir Joseph Fayrer, a senior officer in the Indian Medical Service, told the story of a young, European man of nervous and excitable temperament who was bitten on the cheek by a pet spaniel that was not suspected at all of being rabid.

> It evidently caused him great alarm, and unfortunately he, just at this time, he read a story in *Blackwood's*, entitled 'The Longest Month of My Life'. It was an account of a person who had been bitten by a dog, and who was assured that he was not safe from hydrophobia until thirty days had past. The story made a very deep impression on his mind, and he brooded deeply over it. He destroyed the dog directly

after reading the story. His friends informed me that on the thirtieth day after the bite he began to show symptoms which alarmed them. These rapidly became very grave ... He had no dread of the sight or sound of fluids, and readily tried to drink when told, but was unable to swallow them. Intense horror and consciousness of his condition were present, and he died in a state of exhaustion.[138]

Fayrer wrote that such was the influence of fear and apprehension that he believed people could frighten themselves to death.

In the 1870s alienists, the group of medical practitioners specialising in mental illness, later known was psychiatrists, began to publish on the power of the mind over the body, particularly the influence of emotions. They wrote in the context of the growing incidence of insanity and of degenerationist ideas being adopted in psychiatry.[139] Also, hysteria, great fear, and other forms of extreme emotion were being regarded by alienists as conditions that could be the first step to actual insanity.[140] In 1872, Daniel Tuke, one of Britain's leading experts on madness, published a book entitled *Illustrations of the Influence of the Mind Upon the Body* in 1872.[141] This was the same year that Carl Westphal published his article 'Die Agoraphobie', which is now seen as a classic in developing the modern concept of phobias.[142] Before this date, hydrophobia was the only medically recognised phobia; hence, it was inevitably the initial model for other phobias.[143] Tuke stated that cases of hysterical hydrophobia demonstrated the principle 'of sympathy, or imitation, that is witnessing or reading atrocities could, in individuals lacking moral sense and the control of reason, the images of which are impressed upon the mind through one of the senses'.[144] He went to great pains to distinguish 'real and imaginary rabies', but ended by speculating that 'Fear and Imagination' might have the same effect on 'the higher or ideation centres' as the poisonous virus.[145]

The most prolific writer on the psychological aspects of hydrophobia was W. Lauder Lindsay. He regarded the imagination as another source of spontaneous hydrophobia; indeed, he contented that individual and sporadic collective alarm largely explained the periodic nature of hydrophobia deaths.[146] Thus, hydrophobia scares came and went, in cyclical 'mental epidemics' and he insisted that the panics of 1874, 1876, and 1877, resembled in all essentials episodes in the 1820s and in the late eighteenth century.[147] Lindsay's faith in the mental origins of hydrophobia was strengthened by his view that press reporting of 'murders and suicides tended to multiply such crimes against the body, social or personal'.[148] The minds that were most likely to succumb to hysteria were those that were considered disordered and ill-regulated: illiterate Irishmen, hysterical women, children, and drunkards. For Lindsay, the influence of fear upon disordered minds was most dramatically evident in cases where the victims reproduced the behaviour of the animal that bit them. In Lindsay's own words, 'the imagination begets curious errors in man's simulation of what he conceives to a typical canine or feline hydrophobia'.[149] Hence, barking like a dog or attempting to bite everyone revealed how fear operated on the egregious mind.

In the epizootic of 1876–1877 speculation by doctors on spurious hydrophobia was shaped by the new ideas on anxiety and phobias. Until the 1860s, anxiety states and panics had not been differentiated from other insanities, and hindsight allows us to see that they were mostly absorbed within acute manias.[150] The explicit link between psychological and physical symptoms was first systematically set out by the influential French doctor Auguste Morel, who, in the framework of degenerationist ideas, supposed that disorders of the functional nervous system could influence mind and body, with body affecting mind and vice versa. One condition that was widely discussed was vertigo, which was attributed to panic attacks and disease of the inner ear. As well as identifying specific anxieties, doctors also began to write about general anxiety states, which they sometimes referred to a panophobia – fear of everything. In the longer term it was Freud who developed the idea of anxieties leading to morbid, usually paralysing, physical symptoms, and who also identified many phobic states.[151] However, similar ideas were current around hydrophobia in the 1870s. In letters to medical journals, British doctors speculated about spurious hydrophobic symptoms in terms of 'nervous horror' affecting the functioning of nerves and of 'morbid ideas' creating abnormal nerve forces.[152] The influence of debates on the nature of hydrophobia on the development of medical knowledge of insanity has been missed by historians, perhaps because they have assumed, wrongly it seems, that hydrophobia was always linked to rabies and contagion.

Mental effects made reading the signs and symptoms of hydrophobia very difficult for doctors, but treatment was an even greater challenge. The overwhelming view amongst doctors remained that once the disease had developed nothing could be done other than alleviated suffering. Thomas Dolan typified confidence in early preventive treatment, especially excision and cauterisation; indeed, he claimed that it was the key reason why such a small proportion of mad dog bites led to hydrophobia. Many doctors believed that the virus-poison remained at the site of the bite for many days, and stressed the importance of the victim's constitution in keeping it there and perhaps neutralising it. The struggle between seed and soil explained incubation periods of six months and more. In similar vein, some doctors became optimistic about arresting, if not curing, the disease by controlling symptoms with new sedative drugs such as chloroform and morphine. The benefits of sedation were not only for the patient; they also brought relief to anxious relatives and friends disturbed by witnessing their loved ones seized by violent and incessant convulsions.[153] Not only did chloroform reduce symptoms, it also made the applications of other medicines easier. For example, fear of water and skin sensitivity had made giving medicines orally or by injection difficult. The genre of reports of recoveries from advanced hydrophobia were treated with scepticism in Dolan's volume, as in almost every case it was impossible to demonstrate that the biting dog had been truly rabid and that the patient had not been suffering from spurious hydrophobia.

The second Committee formed in 1877, supported by the Scientific Grants Committee of the British Medical Association, was tied to the Brown Institute in London, and was established to inquire into the causation, pathology, and treatment of rabies and hydrophobia.[154] The Committee consisted of leading veterinarians and doctors, but was loaded towards those with a background in laboratory investigations and commitment of germ theories of infectious disease, including Burdon Sanderson, Ernst Hart (editor of the *British Medical Journal*), Thomas Lauder Brunton, and Dr T. Lauder Callender. The attachment to the Brown Institute, and the presence of Sanderson and Lauder Brunton, who were authors of the infamous *Handbook of Physiology*, attracted the attention of anti-vivisectionists.[155] They speculated that the Committee would produce a 'cage of horrors' and torture animals in the search for a cure.[156] The Committee issued a preliminary report in December 1877, which turned out to their only product, in which they stated that rabies was a dangerous communicable disease, only spread by inoculation.[157] They urged that existing police measures should be enforced, particularly those related to the detention of all ownerless dogs, as 'it is by means of the curs kept in hundreds by persons of the lower class without licenses that the disease is enabled to hold its ground in the metropolitan area'.[158] For the rest of the country, they wanted the uniform application of the dog laws and supported Fleming's calls for public education and information.

The Committee's experimental work was never published. However, there is no doubt that rabies was being assimilated by doctors and veterinarians as a germ disease, caused by a living organism rather than a chemical poison. For example, the delay between bite and disease was explained as being due to the contagious agent being lodged in injured tissues, 'shut up' in a lymph nodule, or rooted elsewhere like a parasite; all of which stopped it entering the blood and reaching the brain and other organs.[159] George Fleming argued that the contagion acted like a ferment in the injured tissues, continually releasing poisons into the blood, and that only when a toxic dose had accumulated did the disease develop.[160] In other words, rabies was produced by a 'double zymosis', first in the bitten part, and afterwards in the system, the result of which was either to multiply the poison, or increase its virulence.[161]

In 1878 the government increased the cost of a dog license from 5s to 7s 6d. The principal aim was to discourage the keeping of the curs that roamed the streets and lanes of the country and hence to remove the main source of rabies. The introduction of licenses in 1867 had brought some success, with the number of licensed dogs rising from 828,320 in 1867 to 1,302,170 in 1876.[162] However, this seemed not to have reduced the number of stray, ownerless, and vagrant dogs. The increase was to try again to deter casual dog ownership. The Chancellor of the Exchequer, Stafford Northcote, told the Commons that he received many letters on the matter from all classes and from across the country for and against.[163] There was a long debate in the Commons on 11 April 1878 where all sides of the what was called 'the dog question' were aired.[164] Most speakers supported the government's aim of reducing the number of useless curs of the poor and 'men who starved their children and beat their wives, but who fed their dogs on new milk and mutton'.[165] However, against such stereotypical views, some MPs, like Charles Parnell the Irish Nationalist, argued that dogs had a humanising effect on their owners and that dog ownership amongst the poor ought to be encouraged.[166] Parnell also argued that 'he did not believe hydrophobia would be diminished one bit ... but that, on the contrary, it would be increased very much; for more stray and mad dogs would roam about the streets'.[167] The exemptions for fox hounds were criticised by Sir George Campbell (Kirkcaldy) as 'an atrocious piece of class legislation'; however, the strong hunting lobby in the Commons ensured these remained, though concessions were won for shepherds and dogs used by blind persons.[168]

Conclusion

In 1879, Ralph Caldecott, the prolific children's illustrator, published a picture book version of Oliver Goldsmith's 1769 poem 'Ode to a Mad

Dog'.[169] (Figure 3.6.) Its cover showed the funeral cortège of an innocent, allegedly mad dog that had been murdered by a mob, surrounded by a phalanx of black dogs in mourning, with their heads bowed.

As noted before, this was a very moral story illustrating the dangers of jumping to conclusions on first impressions and of taking precipitate actions. The book became best seller, appealing to both sentiment and rationality. The poem presented the dog as an innocent, wrongly judged and given summary justice, and ended by encouraging remorse in all

Figure 3.6 'The Mad Dog'

Source: R. Caldecott, *The Mad Dog*, London: Frederick Warne and Co. Ltd, 1879.

humanity. At the same time, the poem suggested that the action had been irrational; those who rushed to kill the dog had not waited for all the facts and had caused a needless death.

This message echoed that being advanced by the growing network of veterinarians, doctors and medical scientists working on rabies and hydrophobia. They were seeking to encourage more rational responses from the public, through better education about the disease, and to persuade government to act more decisively. On the latter they looked for the extension of the set of linked legislative measures that had controlled other contagious animal disease. Indeed, following the lead of Thomas Watson, they hoped that rabies might be extirpated, like the cattle plague and foot-and-mouth disease. At the same time, scientifically minded medical men hoped that laboratory investigations would both confirm the wisdom and practicality of stamping out, by finally demonstrating that rabies was always contagious. This group still sought better means of diagnosis, though doctors were now accepting spurious hydrophobia as a neurotic condition, and were still willing to try new remedies, though with little expectation of success. The enquiries set up by the *Medical Press* and *British Medical Association* had found little that was new or promising, and largely confirmed existing knowledge, or lack of it. However, doctors hoped that the new laboratory medicine might yield insights into both rabies and hydrophobia in the long-term. What they least expected was what they enjoyed in the 1880s from the laboratories of Louis Pasteur – breakthroughs that produced a preventive vaccine for dogs and an effective treatment for humans.

4
Rabies Cured: 'The Millennium of Pasteurism', 1880–1902

The story of Louis Pasteur's achievements with rabies and hydrophobia has become legendary.[1] He began to work on the diseases in December 1880 and reported steady progress to his scientific peers and the public in succeeding years. In the spring of 1884 he announced a vaccine based on a modified rabies virus of reduced virulence, which he claimed to have protected dogs inoculated with the ordinary virus. In the following year he switched from prevention of rabies in dogs to the treatment of hydrophobia in humans; in fact, he aimed to develop a preventive treatment, using the graduated doses of the vaccine to build up immunity during the long incubation period of the disease. The announcement on 26 October 1885 at the Academie des Sciences in Paris that the life of Joseph Meister, an eight-year-old boy from Alsace who had been savagely attacked by a rabid dog, had been saved by the anti-rabies vaccine caused a medical and media sensation.[2] Pasteur subsequently offered his 'cure' free to potential hydrophobia victims from any country and in a matter of weeks his clinic had attracted hundreds of patients from across the world. Such was the profile of this innovation that a public subscription was created to establish a permanent clinic to treat patients and for research to produce vaccines that could cure or prevent other killer diseases. The outcome was the opening in 1889 of the Institut Pasteur in Paris – the world's first purpose built medical research laboratory – which rapidly spawned similar institutes across Europe, North America, and Asia.

This narrative of 'rabies cured', or more accurately 'hydrophobia cured', would have been familiar to contemporaries in Britain in the late 1880s, as it was reported by the national and local press and promoted by the scientific and medical establishment. But not everyone was impressed, Pasteur became the *bête noire* of British antivivisectionists, and it was horror at the 'holocaust of dogs' in his rabies work that sustained the movement in the 1880s.[3] Critics within and outside the medical profession worried that Pasteur was claiming too much, too soon and that he was playing with fire in creating new diseases, such as 'laboratory rabies' in rabbits, a disease that was more uniformly virulent than the familiar 'street rabies' or 'rage des rues'. Indeed, his opponents claimed that Pasteur was creating his own patients, either through the escape of animals from his laboratory or by so worrying dog bite victims that they developed hysterical hydrophobia. Veterinarians were concerned that the focus on the vaccines and the treatment of humans was diverting attention from the control of rabies in dogs and suppressing or stamping out the disease.

Rabies laboratories and 'laboratory rabies'

Two things kept rabies and hydrophobia in the public mind in the early 1880s: first, the continuing relatively high levels of death (see Graph 4.1) and second, the wide circulation of reports from laboratories in France, especially of major advances in Pasteur's laboratory.[4]

As ever, for every death from hydrophobia there were many times the number of mad dog incidents and suspicious dog bites, and newspapers continued to report pandemonium around such events and accounts of agonising deaths.[5] Anxiety levels about the disease ran high in the early 1880s. An example was a letter to the *Lancet* in December 1881, in

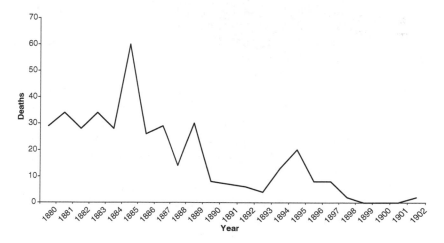

Graph 4.1   Hydrophobia deaths in England and Wales, 1880–1902

which a London doctor, after an unpleasant confrontation with 'a brute' in a patient's home, set out the seven part plan of 'treatment' that he would choose if bitten. He left nothing to chance and stated that he intended to try everything, from vapour baths to curare.[6] There was a perception amongst doctors and veterinarians that Britain was experiencing a new epizootic, and that prevention through licences and muzzling was essential.[7]

Inoculation experiments using the saliva and blood of rabid animals had been an ongoing research programme for many decades, especially in French veterinary schools. In the late 1870s and early 1880s some doctors and veterinarians were suggesting that inoculations of susceptible animals could be used routinely to confirm diagnoses in humans and dogs. An important change was that while veterinarians had previously worked with large animals, such as sheep and dogs, the new work was taking place in laboratories with smaller mammals, such as rabbits and guinea pigs. Such diagnostic methods were being developed in Britain by Thomas Dolan in Halifax, Julius Dreschfeld in Manchester, and William Greenfield at the Brown Institute in London, in the latter part of the investigations of the Hydrophobia Committee of the British Medical Association.[8] However, it was publications in France, by Maurice Raynaud, a Paris surgeon, and Pierre-Victor Galtier, a veterinarian at Lyons, which attracted medical attention and their work first guided Pasteur.

Initially there were two aims to Pasteur's work: to find the best source of infected saliva to develop a vaccine, and to identify the germ of rabies and its properties. In fact, his first novel finding early in 1881 was 'a bacterial organism' in the saliva of humans dying of hydrophobia, which Pasteur cultivated and re-inoculated into rabbits and dogs to produce an entirely new 'saliva germ' disease. Pasteur was read as having shown that rabid dog bites were even more dangerous than previously thought, with the victim liable to disease from up to three germs, those of rabies, pyaemia (blood poisoning), and the 'saliva germ'.[9]

The first claim to have produced immunity to rabies was made by Galtier in 1881. He reported that sheep and goats given the hydrophobia virus intravenously not only did not develop rabies, but also showed immunity when later bitten by a rabid animal. He speculated that the intravenous administration of the virus might be given to animals that were already infected to slow the progress and intensity of the disease; he drew a parallel with the way vaccination could be used, after infection, to moderate smallpox infection.[10] This work was described by Dr Charles Cameron MP, in October 1881 at the Social Science Congress in Dublin, as part of a general review of laboratory research on infectious diseases.[11] Cameron was an enthusiast for bacterial germ theories of disease and reviewed the work of Pasteur, along with that of Galtier, Casimir-Joseph Davaine, Robert Koch, Edward Klebs, Edward Klein, Auguste Chauveau, and others.[12] He concluded with the hope that vaccination would spare future generations the toll of infectious diseases and that it was important that laboratory investigations in comparative pathology should be pursued – an obvious jibe at the activities of antivivisectionists and the restrictions on animal experimentation in Britain.

Cameron's talk did not go unnoticed by antivivisectionists. A letter to their journal the *Zoophilist* immediately complained about the manner in which Pasteur's work was being trumpeted in the press by leading scientists and doctors.[13] An editorial in the journal in March 1882 complained that:

> There is no argument upon which the vivisectionists have relied so confidently, or which they have so persistently brought forward upon every possible occasion as the alleged triumphant success of the inoculation experiments of M. Pasteur and the immense benefits they were (going to) confer upon cattle and their owners.[14]

Antivivisectionists pointed out that Pasteur's work had been attacked by researchers in France and Italy, had been found wanting by a government commission in Hungary, and had been severely criticised by none other than Koch.[15] Over the next couple of years, Pasteur's work received close scrutiny from antivivisectionists, who saw him as 'the assumed demigod of biological science'.[16] They warned about the 'Millennium of Pasteurism', where his unproven results would count for more than the accumulated experience of farmers and veterinarians, and where cruelty to animals was institutionalised. The following description, from *Le Figaro*, of the conditions in which Pasteur kept the dogs used in rabies research was widely quoted:

> Isolated in round and well secured cages are the mad dogs. Some of them are already at the stage of furious madness, biting the bars, devouring hay, and uttering those dismal howls which no one can forget who has once heard them. Other dogs are still in the incubating period and still caressing, with soft eyes, imploring a kind look. [A dog awaiting experiment was] tied down, with wild looks and his body trembling with terror.[17]

In 1884 readers must have relished being told of the problems Pasteur faced trying to expand his laboratories, when local residents complained about the noise from his 'menagerie', their fears about living next to 'Rabies factories', and falling property values.[18] There was wider publicity, for example, the *London Illustrated News* carried an illustrated feature on Pasteur's laboratory in June 1884. (Figure 4.1.)

Also, it did not go unrecognised in Britain that in September 1884 a 'Pasteur Display' was added to the Musée Grevin – Paris's Madame Tussauds. The report on this in the *Zoophilist* noted that, while the catalogue presented his work positively, the 'counterfeit' laboratory had been located in the basement next to the chamber of horrors.[19]

During 1884, reports in scientific and medical journals told a story of Pasteur's progress towards a rabies vaccine, though the assumption was that this would be used to protect dogs.[20] In February 1884, Pasteur announced that he was able to produce immunity experimentally and in May he announced a reliable vaccine that gave predictable and repeatable results. He invited the French Académie des Sciences to

INOCULATED DOG IN CAGE.

Figure 4.1   'Inoculated dog in cage'
Source: *London Illustrated News*, 21 June 1884.

appoint a Special Commission to repeat his experiments and validate his findings. This was a device that he had experienced previously, most famously in his conflict with Pouchet over the spontaneous generation of life in the 1860s, and, as then, there were immediate accusations that the Commission was packed with Pasteur's supporters.[21] Thus, it was no surprise when the Commission confirmed Pasteur's findings in June and September 1884.

In August Pasteur spoke at the International Congress of Hygiene and Demography and summarised his rabies work since 1880. He focused on three points. First, how he had developed a way of producing rabies in the laboratory with great reliability, by inoculating material taken from the brain of a rabid animal, not saliva or blood, directly into the brain of experimental animals. Second, how he had been able to increase and decrease the virulence of the rabies virus by serial passage through different species. Passage through rabbits increased its virulence until the incubation period was a standard six days; he called this the *virus fixeé* and the disease produced was termed 'laboratory rabies' – the more variable natural virus was now said to produce 'street rabies'. Serial passage through monkeys attenuated the virus and it was this weakened virus that Pasteur had used to vaccinate dogs. Third, Pasteur welcomed how the first report of the Commission which had confirmed all his results. Nonetheless, a summary of his talk in the Manchester-based *Medical Chronicle* concluded, 'There still remain important questions to be answered. How long does the protection last? Can a successful protective be practised after the animal has already been bitten by a rabid dog?'[22]

Although not mentioned explicitly by Pasteur himself, the prospect of a post-bite treatment excited press and public attention. This possibility, first raised by Galtier, was mentioned by Pasteur in a talk on 19 May 1884. He reported promising results with monkeys and concluded that 'owing to the long incubation period, I believe that we will be able to render patients resistant with certainty before the disease becomes manifest'.[23] However, he added that this would be some way off as 'proofs had to be collected from different animal species, and almost *ad infinitum*, before human therapeutics can be so bold as to try this mode of prophylaxis on man himself.' Nonetheless, Pasteur was immediately inundated with requests from desperate victims, their families, and their doctors to be allowed the treatment; many had read him as saying that he already had the treatment to hand. According to a member of the North London branch of an antivivisection society, this was typical

Pasteurian hyperbole and 'his third attempt to gain celebrity by such claims'.[24] The *Zoophilist* commented that:

> Pasteur is paying the penalty for his own rashness. He has been deluged with applications, not only from Paris, but from all parts of the world. Letters continue to pour upon him, day after day, from persons who have been bitten, or whose friends have been bitten by dogs.[25]

In succeeding months, Pasteur and the medical press that allegedly 'boomed' his work was attacked by British antivivisectionists for perpetrating a 'cruel delusion' on the public when, on his own admission, many further proofs were needed.[26] Such proofs would mean, of course, further mass cruelty to animals, but it was the raising of public expectations that was seized upon. Pasteur was described as 'a heartless scientist' and likely to be seen as 'amongst the great impostors of the nineteenth century'.[27] J. H. Clarke, a Member of Parliament and active antivivisectionist complained that Pasteur posed as great benefactor of mankind, but as yet had proved nothing.[28] He also argued that Pasteur and other scientists had talked up the threat of hydrophobia, yet it was a very rare disease. A treatment would not save thousands of human lives, rather thousands of animals would be sacrificed for little gain and human lives would be blighted by needless fears of a rare disease.[29]

Hydrophobia cured?

Hydrophobia deaths in Britain totalled 60 in 1885, the third highest on record and double the number notified in the previous year. Predictably, there was great public alarm and considerable medical and veterinary interest in the disease even before the announcement of Pasteur's 'cure' on 26 October. Indeed, in the previous week there had been an exchange of letters in the *Times* about the best means to control street dogs, prompted by hydrophobia deaths in East and South London.[30] Anxiety levels had been raised by an incident in Poplar in July, when a dog had bitten five boys on their way home from school and all died of hydrophobia.[31] The solutions proposed were familiar: the early identification of rabid dogs; the use of the Dog Tax to reduce the number of 'half-starved, ownerless dogs on the streets'; a strict rather than permissive Dogs Act; the rigorous enforcement of muzzling; and the killing of all suspect dogs.[32]

These events meant that the public mind was ready for Pasteur's big day. (Figure 4.2.) The *Times* and other papers reported Pasteur's talk at

Figure 4.2 Louis Pasteur examining the spinal cord of a rabbit, 1885.
Reproduced by permission of the Wellcome Library, London.

the Academie des Sciences in detail, relating its principles, practice, and
the case histories of Joseph Meister and Jean-Baptiste Jupille, the two
boys whose names were to be enshrined in Pasteurian legend.[33] Reports
and editorials eulogised Pasteur: 'a man of genius', who having worked
'so patiently and so wisely to so noble and beneficent an end', had
achieved 'a scientific triumph worth all the conquests in the world'.[34]
Equally admired was his offer to provide the treatment free to anyone
bitten by a rabid dog, a gesture that made him a 'Good Samaritan' and

'benefactor of mankind'. Other papers, like *The Graphic* were more scep-
tical asking, Can we be sure that Joseph Meister was not still incubating
the disease?[35] Nor was the medical press uniformly enthusiastic. An
editorial in the *Lancet* commented that 'M. Pasteur's inferences are san-
guine and premature', while end of the year retrospectives, were still
looking towards controlling rabies by veterinary policing and the Dog
Tax, rather than developing treatment services.[36]

Unsurprisingly given the advance publicity, there was an immediate
response from medicine's critics. Benjamin Bryan, Secretary of the antivivisectionist Victoria Street Society, claimed that Pasteur's earlier work
on anthrax had not only failed, but was also dangerous and could spread
the disease. Dr R. E. Dudgeon, a well-known homoeopath, argued that
Pasteur had only cured two cases and that this proved nothing because
the normal incidence of the disease in those bitten was so low that the
survival of Meister and Jupille was unexceptional. He went on to caution
that both boys were still within the normal range of incubation.[37]
Veterinarians also took advantage of the new profile of rabies, with
George Fleming saying that it was too early to judge Pasteur's treatment
and that however successful it might be, prevention would be better
than cure.[38] Reaffirming the consensus reached between veterinarians
and doctors at the end of 1870s, C. R. Drysdale, an enthusiast for bacteriology and experimental science, also supported stamping out.[39]

Pasteur's 'cure' was strictly speaking a preventive treatment. It was
based on the idea that the incubation period of rabies in humans was so
long that it was possible to induce artificial immunity before the germs
had multiplied sufficiently to produce the disease. The treatment regime
consisted of daily inoculations of weakened rabies germ taken from the
spinal cords of rabbits, not as previously from serial passage through
monkeys.[40] Pasteur produced the vaccine by inoculating his *virus fixée*
into the brain of rabbits and dissecting out the spine after death. The
spine was then suspended in jars in dry, sterile conditions for differing
lengths of time, with Pasteur assuming that attenuated or weakened
rabies germs were produced by exposure to oxygen. Initially he argued
that there were qualitative changes to the virus – it lost virulence; however, he later moved to the view that it was the quantity of active virus
that was reduced. Vaccines were made by macerating spinal tissue in
saline and fresh batches were produced daily, which required a production line of infected rabbits and drying spines. Thus, the weakest vaccine
was made from 15-day-old spines and the strongest from those left for
only five days. Immunity was built up by giving patients successive
inoculations, moving from weak to strong vaccines. Thus, patients

started with two inoculations per day of 15-day-old vaccines, switching after five days to single daily vaccinations, building up so that they were able to 'take' the strongest vaccine after 15 days.

On their first visit to the clinic patients were interrogated over their experiences to determine whether the dog that bit them had been truly rabid. Weeding out the 'worried-well' and possible medical tourists was aimed to avoid the waste of vaccines and meet the objection that Pasteur's success rate was inflated because he was treating many people who had never been exposed to the rabies virus. Once admitted, patients were given a number, before being checked through a turnstile to enter the inoculation room. Patients from all countries, classes, and creeds waited together; it was a point of principle to offer no privileges nor to accept any payment. The vaccines were kept in glasses from which hypodermic syringes were filled, after sterilisation of the needle by passing it through a flame. The patient then was brought forward and instructed to expose their midriff, women being able to go behind a screen to loosen their clothing, and the needle was pushed downwards into a fold of flesh just below the ribs. The inoculations were given by colleagues, such as Joseph Grancher, because Pasteur, not being medically qualified, was unable to treat patients.[41] In the first four months over 500 patients were treated and these came from all over France and the world. Parisians marvelled at the 'most cosmopolitan crowd' that assembled in the garden outside Pasteur's laboratory at 11 o'clock every morning and were surprised that 'there should be so many persons bitten by mad dogs, whereas before Pasteur's discovery there seemed to be so few'.[42] One explanation was that previously 'victims kept their misfortune secret'.[43] Another was that Pasteur was treating many ordinary dog bite victims who had been made anxious by the publicity given to his work; in other words, he was creating patients who would inevitably be 'cured' by his treatment.

Between November 1885 and January 1886, only five dog bite casualties from Britain went to Paris for treatment. The first was a doctor from Oswestry, John Hughes, who was followed by victims from Hertfordshire, London, and Liverpool. The press paid little attention to these patients, only taking up the story when, following the lead of American papers, they reported the fate of four boys from Newark, New Jersey, who were treated in December and returned 'cured' in the new year.[44] Press and medical interest in British victims took off when a group of seven dog bite victims from Bradford arrived in Paris on 14 March.[45] All had been bitten by the same rabid dog on 24 January and had been treated by local doctors, mostly by cauterisation. There had seemingly been no thought of sending them to Pasteur until another casualty, Tom Ashworth, died on 11 March 1886.[46] Questioning revealed that Ashworth, after having had his wounds cauterised, had travelled to Colne in Lancashire, some 23 miles, to be treated by J. R. Hartley, a part-time herbalist, whose 'Hydrophobine' remedy was known and used across northern England.[47]

Ashworth's death revealed that the dog had been truly rabid, and that its other victims were in real peril. The town's Medical Officer of Health (MOH), Thomas Whiteside Hime, an enthusiast for bacteriology and laboratory medicine, telegraphed Pasteur and asked for vaccines to be sent to Bradford. Pasteur turned down the request, saying that the procedure and its materials were too complex, but, of course, he invited all those bitten to attend his clinic. Time was of the essence; it was 49 days since the bites, well into the period when symptoms typically developed. The town's leaders decided to send the victims and to start a subscription to pay for the enterprise. Hime left with his seven charges, four boys, a girl and two men on 13 March and later they were joined by another man. The group became national celebrities and the children were portrayed in *The Graphic* on 3 April. (Figure 4.3.)

Over subsequent days the national and local papers followed the fate of 'Hime's protégés', with detailed reports of their experiences at the clinic.[48] The Bradford papers reported on other British patients and told of mad dog incidents in Peckham and Islington (London), Oldham, Hawes, and Pontefract, as well as in Bradford itself and nearby Keighley.[49] Interestingly, on 13 March the account of the departure of the Bradford group was accompanied by a report from the *New York Times* that the dog that had bitten the New Jersey boys had not been mad after all, so the 'hullabaloo' had been for nothing.[50]

Hime's 'protégés (they were also termed his 'brigade') began treatment on 14 March, the day after a group of around 22 Russian peasants or Moujiks, from Beloi near Smolensk arrived.[51] They had all been bitten by a wolf and their appearance had been eagerly awaited by the press.[52] Indeed, the story of the exotic Russians often took precedence over that of the Bradfordians in the British press. They were good copy. They were 'rough, long-haired … inured to privation' and had not simply suffered bites, rather 'flesh was literally torn off their bodies'; and, of course, they had been bitten by wolves.[53] They had travelled in their normal clothes, but the press dwelt on their national costume as if they had dressed for their new celebrity. Their presence symbolised that something momentous was occurring in Pasteur's clinic. His treatment had attracted over 400 'heterogeneous subjects', of both sexes, of every age and social class,

Figure 4.3 'M. Pasteur's Experiments for the cure of hydrophobia – The doctor and some of his patients'. The illustration shows the five Bradford children, along with a girl from London, and nine Russian patients

Source: The Graphic, 4 April 1886. Reproduced by permission of the Wellcome Library, London.

and from 40 or more countries.[54] The crowd that gathered every morning to receive their 'jabs' was a cosmopolitan, polyglot collection – one correspondent observed that 'Mr Pasteur's waiting-room offered on a small scale the spectacle that must have been witnessed after the failure to build the Tower of Babel'.[55] All were anxious as they might be harbouring not just a fatal disease, but also facing the worst of all possible deaths. Yet, salvation lay in the schedule of inoculations, in the glasses that contained the attenuated virus, and in the expertise of Pasteur and his colleagues. The enterprise became theatre; the inoculations attracted spectators and Parisians seemed to relish the mixture of advanced science, heroic medicine, and suffering humanity.[56] (See Figure 4.4.) A report in the RSPCA journal *Animal World* referred to Pasteur's laboratory and clinic as a 'Temple of Science'.[57]

Figure 4.4 'Rabies vaccination in Pasteur's clinic in Paris', 1887. Dr Grancher is giving the injection as Pasteur looks on. Most illustrations of the clinic show members of the public watching the procedure. Reproduced by permission of the Wellcome Library, London.

The Bradford group also had a distinctive profile. They were the equivalent of the New Jersey boys – the first group of English laboratory cure-seekers.[58] Their leader, Thomas Hime, was fluent in three languages and in modern microbiology, so he mixed widely and acted as a translator on many levels. He helped his own and other patients in the clinic and with their accommodation; he spoke to journalists about Pasteur's methods; and he spent time with the Pasteurians in their vaccine production facility and experimental laboratory. The Bradfordians were portrayed as exotic; a report in the *London Daily News* recorded one Bradfordian's view of the city as, 'Aw cood do wee Pawris un ud canli speock sensible loike', which was translated as 'I could do with Paris, if one could only speak sensible like'.[59]

Back in Bradford a ninth victim came forward, John W. Mitchell, but rather than following the others to Paris, his treatment was organised by antivivisectionists. He was taken to London and given a course of the Buisson vapour bath treatment *gratis*, as Pasteur's opponents sought to replicate the scheme of free treatment.[60] Antivivisectionists claimed that Pasteurism had become a 'craze', and contrasted their natural, tried-and-tested remedy, with happenings in Paris, where 'M. Pasteur is, in fact, experimenting on humans in the dark.'[61]

On their return home 'Hime's protégés' were given a public welcome and their stories filled the local press; interestingly, Mitchell's return received less press attention. When Hime himself returned, he became a missionary for Pasteur, lecturing in the town and soon across the country, gaining a national reputation as an authority on Pasteur's methods and their leading British promoter.[62] In Bradford and the West Riding of Yorkshire it became routine to send mad dog bite victims to Paris. Three victims from Shipley went in May, another from Bradford in June, and Hime himself went again in July 1886 after suffering a wound while experimenting with rabid animals.[63] All of the victims from Bradford and Shipley lived.

The same day that the Bradford group left Paris, one of the Russian patients died; the first of three in the original group to succumb to hydrophobia.[64] This was a blow to Pasteur who up to this time had only experienced one patient dying, Louise Pelletier in December 1885. Her death had been explained by the delay of four weeks between the bite and the first inoculation.[65] This period was, of course, three weeks less than that of the Bradford group when they started their treatment; hence, if one of them had died, no doubt the same explanation would have been used. This situation led many observers to note an asymmetry where Pasteur's group and supporters always claimed success for the treatment if no disease developed, whatever the delay in starting treatment, but always blamed such delays when it failed.[66] Pasteur expressed disappointment at the deaths of the Russian patients, though he suggested that his treatment might have prolonged their lives, and pondered using an 'intensive treatment', with stronger inoculations given earlier to counter the seemingly greater virulence of the virus in wolves.[67] This method required the injection of stronger vaccines in a more concentrated period of time in an attempt to build up immunity more quickly.

The Russian deaths were one factor in the decision of the Local Government Board to appoint an enquiry into Pasteur's methods and treatment.[68] They were prompted by Sir Henry Roscoe, who had been a Professor of Chemistry at Owens College, Manchester and was at this time the Liberal MP for South Manchester. He had built up the Chemistry Department at Owens and in Parliament had taken up the causes of technical education and the promotion of scientific and medical research, becoming a champion of laboratory medicine and a critic of the antivivisectionists. An important factor was that the treatment had begun to come under sustained criticism across Europe, as yet again rabies and hydrophobia divided medical opinion. The Committee had a distinguished membership; it was chaired by Sir James Paget, and included Sir Joseph Lister, John Burdon Sanderson, George Fleming, and Roscoe.[69] However, the key member turned out to be its Secretary, the up and coming Victor Horsley, who was Superintendent at the Brown Institute and Assistant Surgeon at University College Hospital, and who went on to become Britain's leading brain surgeon.[70] The Committee was appointed in April 1886, but did not report until July 1887, which meant that for over a year 'the jury was out' on Pasteur's treatment in Britain. There was intelligence that the report was ready in September 1886, but that it was being held back because of suspect deaths amongst British 'Pasteur patients' linked to the adoption of the 'intensive method' that he had adopted for patients with severe injuries or wolf bites.

Britain's then leading medical authority on the disease, Thomas Dolan, published a critical pamphlet on the Pasteur's treatment in May 1886.[71] First, he argued that Pasteur's statistics of 'cures' were unreliable as it was certain that many of his patients had never been infected, and would never have developed hydrophobia. Second, that most had already been treated by ordinary means, usually cauterisation, and that this would also be a critical factor in their freedom from the disease; indeed, such was his faith in cauterisation that he saw no need for new methods. Third, he argued that Pasteur was inconsistent: he claimed every patient who did not develop the disease as a success, but blamed failures on delays in starting the treatment. Dolan's arguments were immediately taken up by Dr Anna Kingsford, a physician who was also a mystic, a leading Theosophist, and antivivisectionist.[72] She maintained that 'If the Russians who died came too late, then all the Russians came too late'.[73] Pasteur's supporters countered such claims, particularly C. R. Drysdale who engaged in long exchanges with Dolan in the medical and national press over matters like the number of Pasteur patients bitten by truly rabid dogs.[74]

The treatment also came under attack from doctors with Indian experience, who dwelt upon their knowledge of the vagaries of the disease and detailed the many local remedies they had seen in Asia.[75] The most scathing critic was Vincent Richards, the snake poison expert, who wrote in the spring of 1886 that, 'M. Pasteur's method of treatment as far as the world has been enlightened, rest on no firmer basis than that which justifies the vaunted powers of "Holloway's pills" and "Mother Siegel's Soothing Syrup"'.[76] In a letter to the *Lancet* in June 1886 Richards was more measured, simply stating that 'there appears to exist an unreasoning acceptance of assertions in its favour to the exclusion of facts that are unfavourable'.[77] Later, the physician and sanitarian Benjamin Ward Richardson compared Pasteur's vaccines to homeopathic remedies as they treated like-with-like, using the 'isopathic hypothesis of curing a disease by the scientific art of reproducing it and securing its perpetuation'.[78]

After the experience with the Russians in the spring of 1886, Pasteur's response to death of patients had been to surmise that immunity had not been built up quickly enough, especially in patients suffering from severe injuries, from bites to their head and neck, or from wolf bites. In order to try and accelerate the build up of immunity, he gave patients stronger vaccines and at an early time; indeed, they were given one-day-old cords on day 7 and again on day 11, previously the strongest vaccines given had been from five-day-old cords.[79] This method was developed over the summer and became an option for all patients in September and October. In Britain it became associated with two high profile deaths, that of Goffi (Joseph Smith) in London and Arthur Wilde in Sheffield, which followed the much less publicised death in August 1886 of ten-year-old Harry Collinge, from Rawstenstall, Lancashire, the first death of a British patient.[80]

On 20 October Goffi, an attendant at the Brown Institution of all places, died at St Thomas's Hospital. He had been bitten 12 times by a rabid cat at the Brown Institute on 4 September (seemingly not one of Horsley experimental subjects) and had been sent to Paris immediately, starting his treatment the next day.[81] He was given the intensive treatment because of the severity of his wounds and this lasted for an unprecedented 24 days. He returned to London seemingly well and began work again on 10 October. On the 18th he complained of stomach pains and then developed paralysis of his limbs. He died two days later. The signs and symptoms were said by his doctors not to be those of hydrophobia and a post-mortem suggested that he had died of Landry's paralysis.[82] However, Horsley took some spinal matter to test and inoculated rabbits and a dog; surprisingly, all died of paralytic rabies seemingly confirming that Goffi had died of hydrophobia. Critics seized on this, and other similar deaths in Paris, to suggest that the 'intensive treatment' was producing yet another entirely new Pasteurian disease – 'paralytic rabies' in humans.

There is evidence that Pasteur himself was troubled by Goffi's death as it was also sensationalised in the Parisian papers.[83] According to Patrice Debré, Pasteur began 'something akin to a police investigation, including among other things interviews with everyone who had come into contact with the victim'.[84] He learned that Goffi was illegitimate, spoke English and French, could not read, and was a drunk. One night he had fallen in the Seine and after being dragged out suffered fever and vomiting for three days; hence, his long period of treatment might have been due to interruptions due to drunkenness and illness. As well as blaming Goffi's behaviour and morality for the failure of the treatment, Pasteurians turned potential disaster into triumph by claiming to have found that alcoholism creates hypersensitive *terrain* for the rabies virus and was likely to inhibit effective vaccine treatment. Horsley too pursued the case in depth, publishing a lengthy study in 1888.[85]

The second controversial case concerned Arthur Wilde, who died in Rotherham on 3 November. Wilde had been sent to Paris after being bitten by a man called Oates, an alleged hydrophobia patient he was caring for. He too received the intensive treatment for what was certainly an unusual case; he seems to have been the only man-bites-man case ever dealt with by Pasteur. Wilde also returned to work and developed an illness some days later, in his case weakness, dizziness, and breathlessness. The post-mortem found that he died of congestion of the lungs.[86] A report in the local newspaper was headlined, 'Death of a Pasteur Patient Under Suspicious Circumstances', and immediately brought Thomas Whiteside Hime to the Rotherham to investigate.[87] On talking to Wilde's doctors he found no evidence of hydrophobia and was so confident that, unlike Horsley in London, he took no samples for laboratory diagnosis. He wrote to the local papers and the medical press to complain about misreporting and prejudice against Pasteur.[88] However, doubts remained. Hime's claim that Wilde was in fact never bitten by Oates was disputed and antivivisectionists questioned Wilde's mother, who reported that he had suffered pains in the abdomen at the site of the inoculations and had frothed at the mouth.[89] Antivivisectionists and medical critics used the Goffi and Wilde-Oates cases to paint a picture of laboratory scientists creating new diseases and causing rather than curing disease.[90]

The debates on the merits of Pasteur's treatment continued in the medical press. The most important was between clinical and laboratory expertise, respectively Thomas Dolan and Victor Horsley. They exchanged letters in the *British Medical Journal*, with others joining in on both sides, from early September to November 1886.[91] The initial issues were Pasteur's statistics and the principles of the treatment, but their exchanges became increasingly polemical and personal.[92] Horsley ended the correspondence with complaints about misrepresentation, having to answer 'utterly unsupported statements' and accused Dolan of adopting 'the un-English method of fighting'.[93] In December Dolan debated the topic with Thomas Whiteside Hime in Leeds, repeating his by now familiar objections and complaining that Pasteur's supporters in Britain were trying to stifle legitimate criticism.[94] The fact that medical critics were being lumped with antivivisectionists shows that more was at stake than the effectiveness of a single innovation; it seems that for the leaders of the profession the future shape of medicine was at issue. However, Pasteur's supporters were also unhappy. Hime spoke in London at the end of 1886; he ended by complaining that while other countries acted on the matter, by founding Pasteur Institutes, 'The English public were still vainly waiting for the publication of the Royal Commission [*sic*] of inquiry'.[95]

In the first half of 1887 Pasteur's treatment continued to be attacked and defended in the medical press. In January there was major debate in Paris between Joseph Grancher and Michel Peter, which while reported as inconclusive, did nothing to diminish enthusiasm amongst Pasteur's supporters.[96] Reported criticisms increasingly came from abroad, notably from India, where Richards and Gordon continued to doubt the whole laboratory enterprise, and from Germany, where Billroth amongst others assailed the whole of French medical science.[97] Statistics, principles, experimental results, and the 'intensive method' remained the key issues, but there was a new concern – mortality. By the summer of 1887, over 30 patients had died after the treatment, though Pasteurians argued that the figure overstated the true risks as many patients had come too late for effective treatment. Indeed, they celebrated the fact that this was very low given that thousands had by then been treated. But the absolute number allowed the British antivivisectionists to introduce a feature into the *Zoophilist* – 'Pasteur's Hecatomb', which listed those who died after treatment. An enlarged version of this page was sometimes fly-posted around London.[98]

The long awaited report of what can be termed Horsley's Committee was finally published at the end of June 1887.[99] It was surprisingly brief, but gave details of Horsley's own experiments, defended Pasteur's statistics, reviewed the cases of Goffi and Wilde, set out details of how rabies in dogs could be stamped out, and, in an appendix, presented 90 case histories of Pasteur patients, including only one Briton – Thomas Hughes the first British patient in November 1885. As expected, its authors endorsed Pasteur's treatment and defended his use of the 'intensive method', while noting that it had been restricted to severe cases and that the youngest cords now in use were five days old.[100] The antivivisectionists countered with their own French authority Auguste Lutaud, who spoke in Piccadilly on 27 July.[101] What was most surprising was that the authors did not call for a Pasteur Institute to be established in London, despite their estimate that 860 Britons needed the treatment each year.[102] Equally surprising was that the report tacitly recommended that priority be given to stamping out rabies in dogs; once again, prevention was seeming better than cure, no matter how scientifically advanced the latter.[103]

The recommendation to stamp out rabies might have been predicted as in September 1886, Horsley, along with T. H. Huxley, John Tyndall, E. Ray Lankester and the MP John Lubbock, had been a founding member of the Society for the Prevention of Hydrophobia (SPH).[104] The Society's single aim was to persuade the government to introduce the universal muzzling of dogs for six to eight months, possibly a year, to rid the country of rabies once and for all. As well as this specific policy, the Society stood for being 'tough-minded': first, in countering 'hyperfervid and hysterical misstatements' about rabies, and second, in being unsentimental about dogs and muzzling. Indeed, the Society was overtly masculinist, identifying ignorance and emotion with 'Ladies', especially owners of lap dogs, who were said to be 'well-intentioned and sincere, but so as the subject is concerned they simply do not know what they are talking about'.[105] Antivivisection and antimuzzling were inextricably linked on both sides and confirms the gendered character of debates over laboratory medicine and vivisection in late nineteenth century.[106]

Possibly frustrated by the delay with Horsley's Committee, or having advanced warning of its recommendations, in May 1887 the Government appointed a Select Committee on Rabies in Dogs to report on veterinary sanitary measures and possible stamping out.[107] The Committee was asked to avoid the question of Pasteur's treatment, but when Horsley was called as the spokesman of the SPH, the topic inevitably came up. The Select Committee moved with considerable speed; witnesses were interviewed in July and the final report published in mid-August. We discuss the report in the next chapter, but note here the separate committees for hydrophobia in humans and rabies in dogs; indeed, from this point onwards the two issues were on separate trajectories.

The hope of leading medical scientists that the report of Horsley's Committee would bring closure was not met. Over the next three years critics kept up their assault on Pasteur and his vaccine. Thomas Dolan led the attack from within the medical profession, publishing many articles, especially in the *Provincial Medical Journal*, which he edited, and an 1890 popular book *Pasteur and Rabies*.[108] Sceptics with Indian experience continued to counter with anticontagionist views and alternative remedies.[109] And, needless to say, the antivivisectionists kept up their tirades against animal experimentation in general and the hydrophobia treatment in particular. For example, Frances Power Cobbe described Pasteur and his colleagues as 'the motley crew of scientific plotters, and their silly dupes of dog-haters and paragraph spinners, who, between them threaten us with another shameful access of poltroonery and cruelty'.[110] One interesting riposte was made by Horsley in 1889, when he published experimental data on the Buisson treatment, which involved inoculating rabbits with the street virus and then placing them in specially made, rabbit-sized vapour baths![111] Predictably, he found no value in the treatment and equally true to form the antivivisectionists ignored his evidence, except for criticising the cruelty involved.

The Institut Pasteur in Paris was inaugurated on 14 November 1888 and in the early months of 1889 a few politicians and doctors began to call for a British branch. Pasteur Institutes were being proposed for New York, and in Russia and Italy, so Britain seemed to be left behind. The feeling began to gather momentum in March after three children from St Helens in Lancashire, all bitten by the same rabid dog, were sent to Paris.[112] Their trip had been facilitated by Sir Henry Roscoe, whose earlier intervention had brought about the creation of the 'English Commission' and who served as a member. However, Roscoe was instrumental in turning the suggested London Pasteur Institute into a scheme to collect money for a fund to send poor victims to Paris and a gift to Pasteur in recompense of all the Britons he had treated *gratis*. The number of British patients visiting Pasteur's clinic had fallen from

88 in 1886 to 64 in 1887 and 21 in 1888; this was due to the reduced incidence of rabies and publicity given to after-treatment deaths. However, numbers rose in the early months of 1889, a year that saw 130 British patients being treated.[113]

Roscoe's ideas were taken up in May 1889 by the Mayor of London and also enjoyed the support of the Prince of Wales – Queen Victoria's eldest son and heir to the throne. The Prince had visited the Institut Pasteur with his family in June and had donated 100 guineas to the Pasteur Fund.[114] The Lord Mayor called a meeting of the great and the good of metropolitan science and medicine and their political friends on 1 July 1889 to establish a Mansion House Fund. The aim was to collect £5,000 in donations and give half to Pasteur and keep half to fund the expense of the Paris trip for poor victims. After his announcement the Mayor was, in his own words, 'deluged' with letters, pamphlets, and other literature, 'many of them anonymous and a considerable proportion of them scurrilous' from antivivisectionists.[115] This outcry prompted him to go to Paris to see for himself and he returned a convert. When he spoke at the inaugural meeting, the Mayor reassured his audience, no doubt with an eye to the wider public, that Pasteur did not have 'cruel face' and was 'a gentle, humane man'. However, he was made it very clear and very public that the Mansion House Fund was not to support the establishment of a Pasteur Institute in Britain.[116] He claimed this was for business reasons. He argued that on average, between 1886 and 1889, 60 victims per year were going to Paris, of whom he estimated half were too poor to pay for themselves. The costs of travel and accommodation was £25 per head; hence, the total cost for sending 30 poor victims was only £750 per annum – far less than the cost of a new scientific institution. He also argued that 'to establish an institute here would greatly tend to cause delay in the stamping-out process'. He barely mentioned what were undoubtedly the principal reasons – chauvinism and the activities of antivivisectionists. Certain scientists had not given up on idea of establishing a Pasteur Institute in Britain, with Cambridge canvassed as a possible location and Armand Ruffer as Pasteur's chosen man in England.[117] However, the scheme foundered and it was not until 1891 that a similar research institution was created in London. The Mansion House meeting established the fund with much pomp, but by January 1890, only a disappointing £2,839 had been collected, though this had allowed over 30 patients to be sent and eventually some money was sent to Pasteur.[118]

The Saturday following the July meeting at the Mansion House antivivisectionists held their own gathering, presided over by the Bishop of

Ely and addressed by Lawson Tait, the leading surgeon and medical opponent of vivisection.[119] Those present passed resolutions dismissing Pasteur's treatment as misguided and ineffective, but followed the model of the establishment meeting in creating a fund to support poor victims wishing to undergo the Buisson treatment. Perversely, they also attacked leading doctors and scientists for their decision not to establish a Pasteur Institute in London, arguing that it showed they had no faith in the treatment. Nor did it seem that the public had been persuaded. In 1888 there were 14 certified deaths from rabies in Britain, which can be extrapolated to 210 rabid dog bite victims by following the convention of multiplying by 15.[120] Hence, only around 1-in-10 victims went to Paris in 1888.

Estimates on the same basis show that two years later nearly half the victims went to Paris.[121] (See Graph 4.2.) The relative increase was partly due to the availability of funds and partly to the fall in the number of deaths reported; hence, victims felt they had nothing to lose. However, for the majority of sufferers, especially in the north of England and a long way from Dover, Pasteur's treatment remained one of many alternatives, and in most cases the hardest to access. The other options remained what they were half a century earlier: cauterisation and excision by local doctors; traditional remedies from local healers; patent remedies from chemists and druggists; vapour baths, now promoted by antivivisectionists and hydropathists; and of course, self-treatment by sucking the wound, applying caustic, and following various rituals, including the hair of the dog. What shaped choices in particular places was a highly contingent matter, as was the case in East Lancashire and North Cheshire in the early months of 1890, where the complex interaction of local party politics, national science policies, and county police networks led to two groups of victims, bitten by the same dog on the same morning, having very different fates.

Stalybridge, Hyde and Manchester politics

On Saturday, 2 February 1890 the *Stalybridge Reporter* carried a typically melodramatic account of a 'mad dog' incident that had occurred in the town in the previous week.

> On Tuesday morning a few minutes after seven o'clock before daylight had pierced the snow-laden clouds, Constable Millington of Stalybridge borough police force, was on duty in Huddersfield-road beyond Copley, when he was startled out of his meditations and into

an attitude of 'attention' by the sound of a dog approaching him, making a snarling and suspiciously significant sound as if afflicted by the dreaded rabies. The animal, on getting up to the officer, made a spring for his face, but, being on the alert, he warded off the blow with his cape. The dog flew a second time, and the constable was able again to foil it. He then drew his staff, but before he could use it the animal had bolted in the direction of Stalybridge. It seems that just before this a boy named Kane ... had been bitten in the mouth, and when Millington saw the youth he was bleeding from the injury, and ... in a great state of alarm.[122]

The dog then attacked Sergeant Tonge near the Grouse Inn, bit James Saxon nearby, and went into Acres Lane where it 'ran amok at every pedestrian' before biting William Mellor. The dog next appeared in Grosvenor Square, where it bit a postman, Edward Wade, before ending its reign of terror in Vaudrey Street, where two boys delivering milk kept it at bay with stones, before chasing it up Hough Hill and out of town towards Newton and Hyde.

Over an hour later the dog appeared two miles south at Shaw Hall Mills near Hyde, where it bit another postman, Thomas Jackson, on the knee. The *North Cheshire Herald* continued the story.

The dog then made for the main road, and [Jackson] gave chase to it. By the time he reached the cottages the animal was in a most ferocious manner worrying at a little lad named John William Schofield, aged about six years, who was playing in front of his home. The boy's father, William Schofield, who was having his breakfast at the time, hearing a scream, ran out and saw the brute with the little boy on the ground. He ran to the child's assistance immediately, seizing the dog by the throat. Then a desperate struggle ensued between Schofield and the animal, which kept a tenacious hold of the lad, and the father had to prize its lower jaw down to release the child. The dog eventually turned on Schofield and he cried out for assistance. Jackson came upon the scene and he encouraged Schofield to stick to the animal until he went for a gun to Woolley's lower down the road, but she thinking that something was wrong did not open the door. In the meantime an elderly man named Robert Nuttall saw Schofield struggling with the dog, and his efforts to get a poker or some other instrument being fruitless, went to Schofield's assistance regardless of his own safety. Schofield was almost exhausted, for the animal fought desperately to get out of his grasp. Nuttall flung one leg across the hound's back and having clutched it tightly by the throat, he implored Schofield to get out of the way. He struggled hard with the brute, and he maintained a tight grasp until Sgt. Brine, who works at Shaw Hall, went to his residence for his rifle. Nuttall told Brine to take his aim, and when ready he released his hold and pushed the animal forward, and Brine shot the dog through the head.[123]

The *Stalybridge Reporter* was more graphic, claiming that Brine 'riddled its body' with bullets.[124]

The day after the incident, the dead dog, the three Hyde victims and the four from Stalybridge were to be found at their local Town Halls. In Hyde, the Veterinary Inspector, Mr Wardle, and the MOH, George Sidebotham, performed a post-mortem on the dog – the disease was confirmed by the stones and chewed wood in the stomach. Nonetheless, Sidebotham decided to seek confirmation from the Pasteur Institute in Paris by sending the dog's brain and spine.[125] When Jackson and the Schofields reached the Town Hall they were told that funds had been secured to send them to Paris to receive Pasteur's treatment. They left Hyde Junction early that evening, accompanied by Frederick Westmacott, a young doctor from the Manchester Royal Infirmary.[126]

In contrast, the four Stalybridge victims were looked after by the town's Chief Constable William Chadwick, who called in the assistance of his counterpart from the north-east Lancashire town of Clitheroe, John Edwards. The MOH for Stalybridge seems not to have been involved at all and in his Annual Report for 1890 only mentions the dog muzzling order that followed the incident.[127] In the preceding months, Edwards had been championing J. R. Hartley's cure and the double act arrived in Stalybridge on the Wednesday afternoon. All four victims were given the single dose of 'Hydrophobine', a remedy that had never failed.[128]

The impetus to send the Hyde victims to Pasteur came from the town's MOH who had connections with medical science in Manchester through his training and his memberships of the Literary and Philosophical Society and the North West Branch of the Society of Medical Officers of Health.[129] His cousin, John W. Sidebotham, was Hyde's Conservative MP and his brother, Edward J. Sidebotham, was a medical scientist who worked with Sheridan Delépine at the University of Manchester.[130] The trip was allegedly paid for by the generosity of Thomas Ashton, the Mayor of Hyde and owner of the town's largest mill.[131] However, press reports indicated that his son, Thomas Gair Ashton, was the main organiser and some critics maintained that, through Roscoe's contacts, the Mansion House Fund had actually paid. The younger Ashton was the town's prospective Liberal Member of Parliament and he was no doubt advised by Roscoe.[132] Overall, it is likely that rivalry between Conservatives and Liberals, and scientific cooperation facilitated the rapid departures for Paris.

The experiences of the Hyde group were similar to those of other British patients. They were surprised by the size and diversity of the group attending everyday. Jackson wrote home that, 'There are a great many people here: Arabs, French soldiers, Americans, Irish, English, Scotch, all receiving treatment' and that recent arrivals included 'a policeman from London, a Yorkshireman, a Dutch girl and four persons from Leicester'.[133] Every aspect of the experience, from staying in a foreign city to submitting to invasive medical procedures, was novel. They reported many side effects, especially nausea and pains as the patients imagined the vaccine corkscrewing down their legs. Westmacott reported that his cases 'returned home without having experienced discomfort whatever'; however, Jackson told a different story, namely that the 'operation ... is very painful, consisting of lancing on each side of the body'.[134] After their daily treatment, the Hyde group were able to enjoy Paris. Jackson wrote home that it was 'the grandest place in the world', but had the familiar complaints of an English visitor about cold lodgings, inadequate food, and the strange language. Patients tended to stay in designated hotels, as many hoteliers had refused to accommodate Pasteur patients fearing that rabies was contagious person-to-person. While the costs of travel, subsistence and lodgings of the victims were met, families back home relied on local support; for example, Thomas Jackson's wife and daughter lived on money from the Lily of the Valley Pleasure Society, his fellow post office employees, and his friends.[135]

The situation in Stalybridge was quite different. The town with a history of troubled labour relations did not have a civic-minded middle class nor an active MOH. There is no evidence that any consideration was given to following the example of Hyde and other Lancashire towns. Rather, the Chief Constable seized the day and used traditional methods. Thus, he enforced muzzling and ordered strays to be rounded up, seizing 134 dogs in all.[136] He had heard of Hartley's treatment through the lobbying of John Edwards after the successful treatment of children in Clitheroe in 1889. Edwards had written to the Department of Agriculture, suggesting that the government investigate this 'infallible cure' with a view to promoting it nationally for livestock.[137] He then wrote to local councillors across the north of England with glowing reports.

His recommendations were taken and followed up in December 1889 when councillors in Heywood, north of Manchester, called in Hartley after five people had been bitten by a 'mad dog'.[138] In turn, the councillors were so impressed that they issued a memorial, which was again reported in the press and circulated locally. Edwards was so successful with his promotion of Hartley's cure, that it was no longer referred to as the 'Colne Cure' or 'Lancashire Cure', or by its trade name of 'Hydrophobine', but the 'Clitheroe treatment'.[139]

Through the spring and early summer 'Hydrophobine' seemed to have been just as successful as Pasteur's vaccine, none of the January victims died. However, this changed on 19 June when the second Stalybridge victim, James Saxon, became seriously ill and was taken to Ashton-under-Lyne District Infirmary. He died three days later from hydrophobia.[140]

A Coroner's Inquest was held a week later at the New Inn in Ashton. Although convened to confirm the cause of death, the inquest was taken over by Hartley and his supporters and turned into a trial of the 'Clitheroe treatment'. Hartley arrived with a team that included John Edwards and two other supporters: Dr Thomas Jackson from Accrington and Thomas Wood, a solicitor, from Burnley.[141] Unusually, Hartley was allowed to cross-examine witnesses and attempted to vindicate his treatment in three ways. First, he questioned Saxon's mother over whether her son had actually taken his medicine. He suggested that her son had vomited the initial dose and may not have taken the additional dose that he had given him. Hartley had taken this line before in newspaper defences of his treatment, claiming that patients failed to take his Hydrophobine mixture and they did not follow his recommendations to avoid alcohol and adopt a low fat diet. Second, he queried whether the boy's wound had been cauterised. He maintained that his remedy was much less effective after this procedure; he seems to have followed a common assumption that cauterisation drove the poison into the body. Third, Hartley challenged the infirmary doctor – John E. Platt – over the accuracy of the death certification.[142] He exploited the fact that the hospital had sent one of its most junior doctors, who admitted that he had never seen a case before. Hartley argued that Saxon's symptoms were atypical and sounded authoritative – he had records of 500 cases in his book, whereas this was Platt's first. In press correspondence a fourth defence was added by Thomas Wood, that an incubation period of five months was so long that the boy could not actually have died of hydrophobia, or that a partial dose of Hydrophobine might have delayed its onset.[143] The Coroner eventually tired of Hartley's interventions and took back control of the proceedings. The jury confirmed that Saxon died of hydrophobia, despite the Coroner's strange summing up. He steered them towards this verdict, but surprisingly suggested that it may have been due to the bite of a cat or donkey in the intervening months![144] He regretted that the boy had not been sent for vapour bath treatment, but made no mention of Pasteur. The previous January a rabid dog had walked three miles from Stalybridge to Hyde, but it seems that it travelled over two centuries in medical culture, from one still in the eighteenth century to one anticipating the twentieth.

The 'Clitheroe treatment' gained national attention through this episode, but only as a symbol of modern quackery.[145] In 1894 there were further mad dog incidents in Hyde, with seven more victims sent to Paris in August (3), October (2) and November (2). One of those sent in November died of hydrophobia ten months later in September 1895 and although this provoked debate, press reports show that it did not undermine local support for Pasteur's treatment.[146] However, the Hyde death was raised in Parliament, when the Chief Secretary for Ireland was asked by the MP for Cavan if he would now look again at local cures, such as that of Philip Morvan in County Cavan. In his reply George Balfour dismissed the suggestion out of hand, saying that 'We very well remember that some time ago the praises of a Lancashire remedy were very loudly sung, yet, if we mistake not, when put to the test it failed'.[147]

English patients

By the mid-1890s it was commonplace for British, mainly English, rabid dog bite victims to go to Paris for treatment. The numbers increased absolutely and against estimates of total rabid dog bite cases (see Graph 4.2). Most interesting is the percentage of possible patients, which shows a punctuated rise, passing 50 per cent in 1893 and reaching over 90 per cent by the end of the decade. We have considered the dip in 1888 previously, whilst that in the early 1890s was associated with the decline in rabies cases, especially in London and the Home Counties. However, Pasteur still drew patients from across the world, for example over the 1890s there was an increasing number of dog bite victims travelling from India, despite the long lag that necessarily occurred between bite and the start of treatment. Patients from India were first reported separately in 1890, when there were 4, and peaked in 1899 when 62 people were treated, 30 of whom were soldiers.[148] In the following year there were 56, but subsequently all victims based in India were treated at the British Empire's first Pasteur Institute, which was opened in August 1900 at

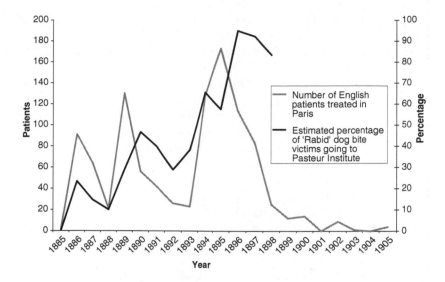

Graph 4.2 Number of English patients treated at the Pasteur Institute in Paris, 1885–1905, and estimated percentage of British rabid dog bite victims treated at the Pasteur Institute, 1885–1902.

Kasauli, a hill station in northern India, under the direction of David Semple. Although the location was remote, 50 miles from Simla, in its first year 331 patients were treated – 146 Europeans and 175 Indians – with only two deaths, both Indian.[149] The institute was given a government grant of Rs 9,500 after calculations showed that there would be saving to the exchequer on the Rs 10,000 being spent each year sending soldiers and officials to Paris.[150]

Why did the Pasteur treatment become the treatment of choice for English patients over the 1890s? First, it became much easier to go. Potential patients found it possible to obtain funding and to find chaperones. How many patients were supported by the Mansion House Fund is unclear, though the lack of evidence of its use suggests that other sources became available, such as local authority support, charitable gifts, private donations, subscriptions and, for the wealthy, self-funding. Scientifically minded doctors were only too willingly to accompany patients and local worthies also took on the duty, as in 1895 when the vicar of Sunbury, F. H. Vigne, accompanied 11 children from the local school.[151] There were also local rivalries to consider, with doctors and civic leaders wanting to be seen to be providing state-of-the-art care and charity, and hydrophobia was particularly useful as children were being saved in melodramatic circumstances.

Second, one way that the orthodox medical profession had used its growing power was to discredit alternative remedies and marginalise its practitioners; hence, by the final decade of the nineteenth century there were fewer alternative healers in the medical marketplace. For example, Hartley was singled out for vilification in national journals.[152] This meant that apart from cauterisation, which was still recommended for all dog bites, the choice was between Pasteur or Buisson. Doctors also argued that the stay in Paris could reassure victims who were anxious or of a nervous disposition, or at the very least it would divert and occupy their mind. Third, the evidence presented by the doctors at the Institut Pasteur showed that the death rate for all patients treated was less than 1 per cent. Moreover, those who died after the treatment were mostly patients with the severest bites, with bites to the head, or with the longest delay between bite and the start of treatment. If the treatment did no harm, then patients had nothing to fear. Antivivisectionists produced different numbers with many more deaths – they aggregated deaths from all the Pasteur Institutes across the world. However, antivivisectionism was in decline in terms of its public profile and was weakest in the north of England where most deaths had occurred.[153] When campaigning opportunities arose in London, as with the death of Ethel Wilkins in November 1896, antivivisectionists found few friends. For example, the Rev. Vigne wrote to the *Lancet* about their tactics with parents in his parish.

> Incredible as it may appear, yet it is still a fact, that someone has circulated broadcast in Sunbury (one of the parents had five copies left at his door) a number of the Star containing harrowing details of Ethel Wilkins' death &c. ... One poor woman (her husband was at night work) I am told paced the road near her house for hours in unutterable anguish. What is this but refined cruelty? If the anti-Pasteurians wish for a hearing, such methods as this is the wrong way to get it; it must, indeed, be a weak case which needs to be urged in this way.[154]

After Ethel Wilkins only three more English patients died following treatment in Paris.

The 1890s also saw narratives by former Pasteur patients being published in popular journals, many of which told of experiences in the late-1880s.[155] Previously this genre had been dominated by accounts written by doctors, but these were replaced by lay narratives. Amongst the most interesting was one by Olga Beatty Kingston, published in *Woman at Home* in 1894, and telling of her experiences back in 1889 when she was a child. A fortnight after being bitten by a dog and seemingly unconnected with this, her father took her on holiday to Paris to see the sights. He only revealed the true reason for the visit when their coach neared the Rue Dutot. The author remembered that:

[M]y father, as though suddenly inspired by a brilliant thought ... exclaimed, 'By the way, child, I believe there is a clever old gentleman lives in this neighbourhood who knows how to treat people who have been bitten by dogs. He has a large building all to himself, where people consult him, and besides, he keeps quite a private zoological garden (presumably for the amusement of his patients), the occupants of which chiefly consist of rabbits, guinea pigs, rats, and dogs. Now, I think it wouldn't be a bad idea to pay him a visit, and show him your hand, eh?[156]

The author's anxieties were short-lived, in part because of her father's deception and in part because she was seemingly charmed by Pasteur. He joked with her and she wrote he had 'that kindest of faces, furrowed with time and care, yet wearing a gentle and sweet expression. ... His beard was grizzled, his eye had a pleasant twinkle ... and [he had a] cordial and genial manner.[157] Before she left Pasteur gave her an autographed photo and she continued to be a fan, writing on a number of occasions and receiving another signed photograph in 1893. Other reminiscences were less flattering to Pasteur, his institute and the treatment, but all were thankful that their lives had been saved, and told a story that recommended the treatment to other rabid dog bite victims.

Conclusion

Pasteur and his supporters changed rabies and hydrophobia, literally and metaphorically. Work in his laboratory produced several 'new diseases': the 'saliva germ' disease, laboratory rabies, paralytic hydrophobia in humans, and the ordinary disease was remade as 'street rabies' (rage des rues). Pasteurian researchers also disciplined the disease, bringing the diverse phenomena of the 'field' into the controlled environment of the laboratory, where they were able to manipulate the virus in animal bodies and in flasks, and to use it in the clinic to stimulate patients' immunity in controlled ways. These developments, almost all produced in Paris, but also repeated and developed at other sites, changed the medical profile of the disease internationally, moving it from the margins of medicine and veterinary practice to become iconic for both professions. In the late 1880s, rabies research was at the forefront of medical progress; it exemplified the value of laboratory researches, and brought the notion of modern medical breakthroughs to the public sphere. All this changed the public profile of hydrophobia, so much so that Pasteur's critics in France and elsewhere accused him of producing panics, anxiety, and the over-reporting of the disease. Critics everywhere suggested that the 'booming' of the vaccine directly benefited his statistics, as many of his patients were likely to have been bitten by non-rabid dogs or be nervous types. Antivivisectionists in Britain were more direct; they said that he falsified his results and killed his patients. Overall, it was not just Pasteur's hydrophobia treatment that was contested, but the whole nature and meaning of the disease, its causes, symptoms, pathology, treatment, and for some, its very reality. After Pasteur, rabies and hydrophobia were new diseases and new cultural phenomena.

Was there a 'millennium of Pasteurism'? Strictly speaking, even a century on, it is too early to say. However, there is no doubt that the origins of all modern vaccines lie in the work undertaken by Pasteur and his colleagues from the late 1870s and there is equally no doubt that vaccines have since saved tens of millions of lives. Pasteur's vaccines did open a new era in medicine. More modestly then, Were the final decades of the nineteenth century an 'era of Pasteurism'? The answer must be yes and this even for Britain, the country that kept rabies treatment at arms length and chose not to establish a Pasteur Institute. There is no doubt that Pasteur and his work on hydrophobia and rabies was iconic for medicine; he was not simply the leading international medical scientist of the day, but someone who had invented that role and made it the epitome of medicine.[158] This status predated his rabies work as he had been lionised at the International Medical Congress in London in 1881 for his work on disease-germs and preventive vaccines for animal diseases, but it was his 'breakthrough' with hydrophobia that brought him wider public attention and symbolised the power of the new laboratory medicine.

5
Rabies Banished: Muzzling and Its Discontents, 1885–1902

None other than Louis Pasteur recommended that the authorities in Britain should concentrate their efforts on eradicating rabies rather than establishing a treatment service.[1] This goal had been advocated for many decades, and since September 1886 it had been actively pursued by through the Society for the Prevention of Hydrophobia and the Reform of the Dog Laws (SPH).[2] While putatively established to combat the rising number of deaths from hydrophobia, the Society was in fact a direct response to the creation of the Dog Owners' Protection Association (DOPA) – 'the Society of the anti-muzzlers' – in August 1886.[3] This set the stage for a protracted struggle between, according to their opponents, 'the "hydrophobists" and advocates of a cruel and unchristian practice – muzzling', and 'unwise dog lovers, who care more for the life of "that divine animal" than for the lives of human beings'.[4] The two sides continued to fight at every level for over a decade, from attempts to amend legislation in Parliament to direct action on the street. Indeed, DOPA was formed as the result of the 'Baker Street affair', when, in June 1886, Miss Frances Ravell was arrested for pouring water over the head of a police inspector who had ordered her unmuzzled dog be killed on her doorstep. The two sides were never reconciled, and even when rabies disappeared in the early 1900s, they still argued over whether this had been due to muzzling or to other factors.

In this chapter we chart attempts to suppress rabies in dogs from the epizootic of 1885, which coincided with the announcement of Pasteur's treatment, through to the claim by the Board of Agriculture in 1902 that rabies had been eradicated from Britain. The notifications for hydrophobia deaths and reported rabies cases are set out in Graph 5.1. The numbers are presented on a logarithmic scale to capture the different magnitudes, and it is worth commenting that the similarity in the shape of the

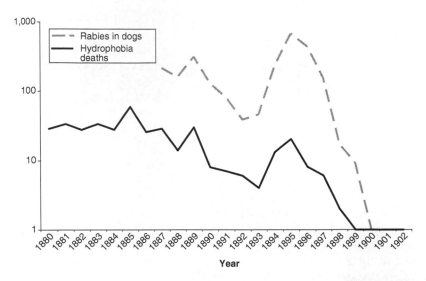

Graph 5.1   Number of reported cases of rabies in dogs, 1887–1902, and hydropho-bia deaths in humans in England and Wales, 1880–1902

Logarithmic graphs do not allow zero on the y-axis, only 1. The actual number of rabies incidents in 1901, and 1903 was in fact zero, and hydrophobia deaths in 1899, 1900, 1901, and 1903 were also zero. In 1902 there were two deaths from hydrophobia, one an imported case and the other a 'suspicious case' of a man alleged to have been bitten two years previously. Neither was related to any of the rabies incidents reported in that year, and the Board of Agriculture stated that they did not damage their claims over the cessation of deaths from hydrophobia in Britain.

curves for 1887–1902 suggests that more or less rabies produced more or less hydrophobia.[5]

The epizootic of 1885–1886

Notified deaths from hydrophobia doubled between 1884 and 1885 to 60.[6] The return of the disease to London, which saw 21 deaths in 1885, ensured the outbreak was well reported in the national press and pro-fessional journals, and discussed in Parliament.[7] Matters came to a head in October, coincidentally the month of Pasteur's 'breakthrough', when the coroner for West Middlesex, at the end of yet another inquest confirming an unfortunate death from hydrophobia, called for stricter controls on dogs.[8] The author Rider Haggard worried about his three children on the streets of the capital where 'mad dogs ... swarm'; he was moved to suggest that 'it would be better to destroy or banish every dog in London, rather than allow one more human being to perish'.[9] Such pleas were echoed in the medical press and by veterinarians in the *Times*; Thomas Price called for better controls of dog ownership and George Fleming once again advanced stamping out.[10] On 20 November, Sir Edmund Henderson, the Commissioner of the Metropolitan Police, issued an Order under the Metropolitan Police Act, 1867, for all dogs on the street to be muzzled or led, with the associated warning that all strays would be seized, taken to the Dogs' Home, and killed if not claimed. (See Figure 5.1.) In the first nine months of the year around 1,500 dogs were seized by the police in London each month; the number rose to 4,950 in November and 9,189 in December.[11] Many owners lost valued pets and others were fined, though some magistrates complained that they did not have the powers to apply penalties, which led Henderson to issue a second Order under the Dogs Act, 1871, that allowed fines. The complexities of the legislation, designed principally for other purposes, meant that one Act was needed to enable muzzling and the other for penalties.[12]

The Royal Society for the Protection of Animals (RSPCA) supported muzzling and the killing of strays as the lesser of two evils, the greater one being the suffering that rabies caused to dogs and humans. However, an editorial in their journal, *Animal World*, warned that muzzling was not a foolproof method.[13] The vast majority of rabid dogs would be unmuzzled and the muzzled dog would be defenceless against such a foe and hence more likely to be infected. The saving grace was that the muzzle acted as a badge, indicating a well-cared-for dog and responsible owner, allowing the police to concentrate on the unmuzzled strays of thoughtless owners. Second, the editorial argued that the muzzle was taken off at home, and that it was here that rabid behaviour was most likely to show. So, either the careful owner would be at risk or the slyness and shyness that were early symptoms would lead a dog to slink away to haunt the streets. Third, the writer worried that as muzzles were painful and many were thrown off, or made dogs agitated and aggressive, many innocent animals would be suspected of 'madness' and wrongly slain. The piece concluded that such dangers and suffering were only justified if rabies was finally controlled. This view was shown to have wide assent amongst members, for when George Fleming spoke at the Annual General Meeting of the Society in August 1886, he was cheered when he expressed his pro-muzzling views.[14]

Known then, as now, as an 'establishment' organisation, it is unsurprising to find the RSPCA following government policy and professional advice.[15] Again, then as now, there were more radical animal welfare

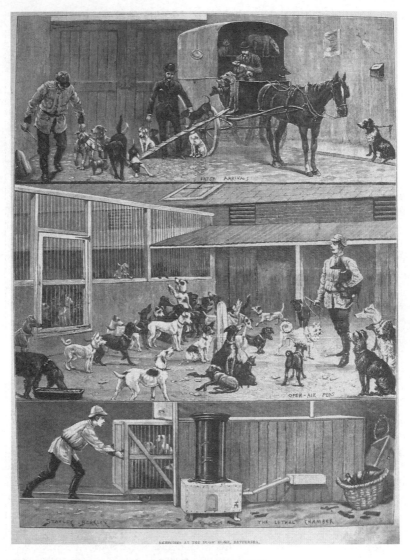

Figure 5.1 'Sketches at the Dogs' Home, Battersea'
Source: *Illustrated London News*, 1886.

groups, and pro-muzzlers like Fleming expected to 'have to endure the howlings or other dolorous utterances of the dog worshippers ... who savagely resent the least insult to their idols'.[16] They were not disappointed, not least when Miss Ravell drenched the police inspector in Baker Street on 14 June. Her dog, which had been let out muzzled by a servant, was reported to the police for its strange behaviour and was found having thrown off its muzzle. Three policemen attended and Inspector Prendergast ordered that the dog be immediately killed by one of the constables. Miss Ravell tried to stop the killing from her first floor window, taking direct action after her pleas went unheeded. She was summoned for assaulting a police officer, but brought a counter summons for malicious cruelty to her dog.[17] The judge dismissed the case against the police, found her guilty as charged and made her pay a fine and costs.[18] After this episode, Queen Victoria was rumoured to have written to the Chief Commissioner to ask if such brutal methods were necessary.

Over the summer of 1886, the London courts were full of cases where owners were fined £2–3 under the Dogs Act.[19] An increasing number of owners in and around London drew legal and financial support from DOPA, also known as the 'Dog Defence Fund', which was first led by Professor Woodruffe Hill, a former fellow of the Royal Veterinary College and lecturer at the College of Agriculture, and by George Candy, QC.[20] As well as supporting individuals, DOPA challenged the whole basis of the court actions under the Dogs Act, 1871. The objections were taken to the High Court, with the plea that 'control' should not only mean a dog being muzzled or led, but could also refer to a quiet, well-trained and well-behaved dog with a responsible owner. The case was lost, but it set the scene for a decade of court battles over dogs and the law.[21]

Whilst supporting muzzling in principle, the RSPCA was critical of the manner in which the police were enforcing the regulations. At the Annual Meeting that cheered Fleming, there were objections to 'uninstructed policeman' being left to determine whether a dog was mad, and to the summary manner in which they went round 'knocking the brains out' of dogs.[22] In 1886, the streets of London were volatile and the public was sensitive to heavy-handed police actions. For example, February had seen the 'Black Monday riot' in Trafalgar Square, where police and trade unionists clashed, which led to the resignation of the Chief Commissioner. Officials at the Kennel Club also supported muzzling, but worried that pro- and anti- factions were becoming too partisan. On events in the street, they felt that implementation of muzzling was becoming arbitrary.[23] In some districts no time limit was specified by local authorities, while in others restrictions were off and on, as in Nottingham. Also, there remained the problem that because of the problem of local authority boundaries, a dog might have to be muzzled on one side of a street but not the other.

The incidence of reported rabies diminished in London during 1886 and the police were said to have tired of dog-catching and the squabbles in court; seemingly it was doing little for police morale or public relations given the other tensions on the streets. Hence, in December the new Commissioner, Sir Charles Warren, allowed the Orders to lapse. However, by then the control options had increased as rabies had been added to the Contagious Diseases (Animals) Act (CD(A)A), which was administered by the Privy Council. Indeed, a new Rabies Order had come into force on 1 October. At last Fleming's lobbying had been successful, though it was somewhat of an anomaly in legislation designed for economic animals. The Order covered England, Wales, and Scotland, and was designed to address the continuing problems outside London, for example in Nottingham.[24] There were technical faults in the Order, no doubt spotted by members of DOPA, which led to it being revoked, revised, and reissued with effect from 28 February 1887. The first Order had given local authorities powers 'For providing for the muzzling of dogs' and 'For providing for the keeping of dogs under control'.[25] The second Order gave the police and magistrates less discretion, giving powers 'For providing for the muzzling of dogs *except when securely attached to or kept within a kennel, house, building or other like place*' and 'For providing for the keeping of dogs under control (*otherwise than by muzzling*)' – changes in italics. These concessions seemed to encourage the anti-muzzlers, who now claimed that the decline in rabies since 1885 had been due, not to muzzling as such, but to the reduction in the number of strays on the street and to greater public vigilance in the face of the publicity given to the epizootic. Organisations such as DOPA, and individuals like George Jesse, now pressed for stricter registration and for dogs to wear identification badges on a collar as an alternative to muzzling. Their aim was to focus on legislation that was enforceable and was not cruel to dogs.[26] Their alternative was to focus on dog owners not dogs, and to try to ensure that all owners were responsible and caring. In other words, they hoped that registration would either, deter ownership amongst the poor and ignorant, or facilitate the identification and removal of the 'uncared for, vagabond animals which are the true source of that exceedingly rare disease – rabies'.[27]

The preparation of a Muzzling Bill by the Victor Horsley and SPH, along with the parallel uncertainties of the Pasteur treatment and the delay in the report of Horsley's Committee, led the government to appoint a Select Committee on Rabies in Dogs on 17 May.[28] (Figure 5.2.) As noted in the Chapter 4, in taking evidence and in their final report the Committee were unable to ignore 'M. Pasteur's system'. The Committee questioned established authorities such as Horsley, Fleming, Hunting, Sewell and Brown, who all supported muzzling, and who were

Figure 5.2 Victor Horsley with his fox terrier, c1885.
Reproduced by permission of the Wellcome Library,
London.

styled by their opponents as 'the muzzle men'. However, it was also
heard from a number of doctors and veterinarians who sided with the
anti-muzzlers. As we showed in the Chapter 4, many of this group were
vocal opponents of Pasteur's treatment, such as Charles Taylor Bell, C. A.
Gordon, and William Whitmarsh, and a key part of their case against
the muzzle was their rejection of germ theories and continuing belief that
rabies could arise spontaneously.[29] Thus, Woodruffe Hill stated that rabies
came from inbreeding which had debilitated the nervous system of dogs
and bitches, especially foxhounds and black and tans.[30] Queen Victoria's
veterinary surgeon, Charles Rotherham, maintained that rabies was a
hereditary affliction and that the irritation from a muzzle could bring out
dormant disease. Hence, just as the antivivisectionist opponents of

Pasteur had their alternative ideas on treatment, so did the anti-muzzlers; indeed, there was considerable overlap in membership and opinion between the two groups.[31] For example, Woodruffe Hill later wrote that 1887 was memorable for the 'intense public excitement' over rabies and argued that the 'morbid minds of the vivisectors may have done long-term damage to the human–dog bond'. He want on that they had 'magnified the complaint and painted harmless affections in ludicrous colours. These morbid minds being suffered to run riot, worked an almost irreparable injury on our ordained companions and most devoted friends'.[32]

The Select Committee worked speedily and published its report in July 1887. Overall it was received as even-handed, steering a middle course between the two sides. Generally the report sided with the leading doctors and veterinarians who argued from both experience and the new laboratory methods, that rabies was always contagious and could be controlled by halting transmission. Hence, it recommended the continued use of muzzles, though qualified this by suggesting that enforcement was restricted to times when rabies was present in a district.[33] The reasoning was that public opinion was so hostile to muzzling, seizures, killing, and the fining of dog owners, that measures could only be enforced when rabies was 'abnormally prevalent', though there was, of course, controversy about assessing prevalence.[34] A further concession to the anti-muzzlers was the recommendation that Privy Council Orders should be widened to include strays and ownerless dogs, and not just apply to obviously rabid animals, and that 'local authorities should have the power to order that dogs should wear badges which may identify their owner'. The Committee made no recommendation on the vexed question of uniform national muzzling versus variable local measures. However, in the discussion of the issue, and following the precedent of much of nineteenth-century legislation that was 'enabling' and left implementation to local initiative, the report stated that such were 'great advantages in ensuring active cooperation of local opinion, that it seems undesirable, except in some pressing emergency, to enforce rather than allow'.[35] One such instance envisaged was when 'one local authority is crippled by the inertness or opposition of its neighbours', in which case the Privy Council might extend the control area.

To muzzle or not to muzzle? Chaplin and 'the Official Plan'

After the inclusion of rabies in the CD(A)A, incidents had to be reported and recorded. Between 1887 and 1888 they fell, but doubled in the following year (see Graph 4.2). Over half the new rabies incidents were in London and surrounding counties, and in June 1889 the London County Council (LCC) came under pressure to act. In the event, they passed the buck to the Privy Council asking that they implement uniform measures over the whole country. In the event, the Privy Council chose only to implement muzzling orders in the metropolitan area, stating that they had been forced to step in as the LCC had 'failed to execute and enforce' appropriate measures.[36] The chief travelling veterinary inspector, Major John Tennant, was put in charge, though he had to work with and through the police. On 10 July, just over a week after the Mansion House Meeting to honour Pasteur, London was 'placarded' with notices headlined: 'MUZZLING DOGS' and 'SEIZURE AND DETENTION OF STRAY AND OF DOGS NOT MUZZLED'.[37] That month the police collected 1,457 stray and killed 58 dogs on the street. The Order antagonised both sides of the muzzling issue: the pro-muzzlers were unhappy that it was local rather than national, while the anti-muzzlers once again objected to a cruel and ineffective policy. However, the debate entered new territory in two areas; first, looking to and learning the lessons of rabies control in other countries, and second, debating the design and effectiveness of different types of muzzle.

The books on rabies and hydrophobia published by Fleming, Dolan and others in the 1870s and 1880s had chapters on the experience of other countries and their control measures. The Select Committee had received detailed accounts of the measures taken in many countries, but was particularly interested in Berlin where muzzling had been adopted and rabies had been declared extinct in the early 1880s.[38] There were other instances, like Baden and Bavaria, where registration seemed to have effected eradication. Indeed, everyone was able to find a Continental exemplar of their preferred policy. In his evidence, John Colam, Secretary of the RSPCA and a leading figure at Battersea Dogs' Home, had made much of European opinion against muzzling.[39] In July 1889 an article in the *Kennel Gazette* had reviewed the three Continental models for a rabies-free country: registration (Baden); quarantines (Sweden), and muzzling (Berlin), concluding that muzzling would only be needed in Britain at times when rabies was epizootic.[40] When muzzling was deemed necessary, a key question was what type to use and the balance between restricting the ability of a dog to bite and cruelty.[41] There were two types: the older leather strap simply drawn round the dog's nose and mouth, and the new wire models that, according to their supporters, could be adapted to any breed.[42]

Leather straps were widely used and widely condemned; if tied tightly, they prevented the dog from drinking, panting and breathing properly, if left loose they were ineffective and could be easily thrown off. Owners of bulldogs and other short nose breeds had long complained that muzzling orders discriminated against them as it was impossible to muzzle such dogs, though the dog journals contained recommendations of suppliers for all breeds. The alleged advantages of wire muzzles were that dogs found them harder to remove and they allowed the dog to breathe and drink, though critics questioned whether this was so.[43] And there remained the problem of owners; even if a dog accepted a muzzle, could owners be trusted to fit it properly?

The objections to muzzling went beyond the technical issues to questions of sentiment in the new age. In supporting the Order, an editorial in the *Times* approvingly quoted the neurologist W. R. Gowers's condemnation of anti-muzzlers as those whose 'perverted sentiment ought to be held in universal abhorrence as a disgrace to humanity'.[44] Horsley and Roscoe were aggravating the situation by canvassing for a Parliamentary Bill to impose universal muzzling until rabies was extinct, which their opponents decided meant muzzling for twenty years![45] Their old antivivisectionist foe, Frances Power Cobbe, joined with the opponents and wrote in typical vein to the *Times* in October 1889,

> My objection to muzzles is that they are teaching the British public to regard with suspicion, dread, and finally hatred animals whose attachment to mankind has been a source of pure and humanizing pleasure to millions, and which has formed a link (surely not undersigned by the Creator?) between our race and all other tribes of earth and air.[46]

She went on to suggest that four out of five people dying from hydrophobia did so from 'nervous hydrophobia, neither more nor less than the absolute result of fear'. A correspondent to *Animal World* in February 1890 speculated as follows:

> Thinking over the craze for muzzling and the terror of hydrophobia ... I can but explain it to myself in one way; viz., that it is the outcome of the nervous, overstrained temperament of the present day; our nerves are simply breaking down, and we are getting afraid of every shadow.[47]

Democracy was also at stake. The anti-muzzlers claimed that their views and policies were not receiving a fair hearing because leading scientists and veterinarians had captured the ear of the political elite. George Candy compared the two sides as follows: one side looks on the muzzle as 'the abomination of desolation', the other as 'the symbol of a millennial condition of society'.[48]

The leadership of the RSPCA was divided on the issue, their journal carried opinion for and against. The leading critic was John Colam, the Society's Secretary for over 30 years, who objected to muzzling on the standard grounds: that muzzled dogs, because they had responsible owners never, developed rabies; that these 'respectable dogs' were rendered defenceless against other, possibly rabid, dogs; that muzzles frightened the public and irritated dogs; and that dogs were most dangerous at home when unmuzzled. Also, he suggested that statistics from both home and abroad told against the effectiveness of the practice, and that there were too many exemptions for sporting dogs, which were probably a major source of the disease. Against this Everett Millais typified those members of the Society who argued that the short-term suffering of dogs and owners, during the twelve-month period of universal muzzling necessary to eradicate the disease, was worth the long-term gain of a rabies-free country. In the vacuum created by the ambivalence of the Society, in December 1889, Woodruffe Hill reconstituted DOPA as The Metropolitan Canine Defence and Benevolent Institute – which became the National Canine Defence League (NCDL) in 1892. Its aims were 'to shield our canine companions from all forms of cruelty – notably, those that arise from time to time through the ignorance and thoughtlessness of *Alarmists* and *Hydrophobic Panic Creators*', and 'arbitrary police measures ... to render canine existence more endurable, and the human mind less disturbed and agitated' (italics in original).[49]

Woodruffe Hill was also responding to the extension in December 1889 of the Muzzling Order beyond London to surrounding counties, which further raised the political stakes. This decision was taken by the newly formed Board of Agriculture which had taken over rabies policy at its formation in September. Its first President was Henry Chaplin, a high Tory from Lincolnshire who 'was every inch the squire'.[50] He was a renowned sportsman, having owned the winner of the 1867 Derby and was a keen huntsman with his own pack, first the Burton and when that was divided, the Blakeney. He was lobbied over muzzling through the autumn and in December received two deputations on the same day: one led by Horsley from the SPH asking for uniform, national muzzling, the other led by Woodruffe Hill from DOPA asking for the end of muzzling and the introduction of registration.[51] Chaplin took the middle ground and would probably have agreed with the writer of an article in *Fairplay* who observed:

> There is a great deal of false sentiment on one side in this matter, and a great deal of pig-headedness on the other. People write and gush about their "four-footed friends" as if the lives of all the dogs in Christendom could, or ought to be weighted against the life of one man, woman or child ... . On the other hand, a working union of theorists and red-tapists lead to widely loop-holed regulations utterly useless for a preventing the disease, but of infinite capacity for provoking discontent and annoyance.[52]

Horsley was spoiling for a fight and inflamed affairs when he hijacked a meeting of DOPA a few weeks later.[53] He arrived with a group of medical students who heckled the speakers. Then he proposed a motion in favour of muzzling, which was apparently won because he had packed the meeting with his student supporters. The anti-muzzlers were outraged and contrasted his behaviour with the closed ticket meeting at the Mansion House that had celebrated Pasteur.[54] One wrote that vivisection had turned Horsley's mind, 'Dr Horsley is simply hysterical, the very natural result of his methods of research; he had overdone his nerves, and like all hysterical people, regards those who look calmly at what fills him with terror as blind and obstinate fools.'[55]

Chaplin's extension of the Rabies Order to the Home Counties shocked the anti-muzzlers who had hoped that a country-gentleman Tory, who owned hounds, would side with them; instead they had to confront 'the blunders and stupidity of the Minister of Agriculture'.[56] Chaplin explained his decision to a deputation from Kent in January 1890 as following the recommendations of the 1887 Select Committee and the lessons of London in 1886–1887, where muzzling had reduced rabies, and Nottingham in 1886, where failure to sustain controls had led to its re-emergence.[57] He continued to disappoint the SPH by stating his reluctance to introduce the 'extreme' measure of universal muzzling, because it was unnecessary and would not have had public support.[58] However, from the perspective of the leaders of towns in Kent like Maidstone, where there had been no rabies, the County Order was experienced as 'universal' and extreme.[59]

Anger spilled over into the press. In London, the *Standard* was the main outlet for protests and in January its columns were full of letters from Kent pointing out the futility of compulsory muzzling orders. Correspondents argued that there was no rabies in the county, that muzzling would not prevent the spread of rabies, and that really dangerous dogs remained uncontrolled.[60] A new element in the south-east was that some Tories were so affronted by 'the present silly, aggravating and utterly useless' Orders that they threatened to vote against the government, while others said they would leave the party.[61] In Manchester the papers were inundated with letters repeating the claim that Muzzling Orders were 'base, cruel, ignorant and barbarous' and calling for their replacement with a registration scheme.[62] On 3 February the Mayor was petitioned to suspend the local scheme and lobby the Minister for registration.[63] In Sheffield the Mayor also announced that he would petition for exemption, in part because any measures would be useless because adjoining Derbyshire was not included. However, most correspondents in the local press were pro-muzzling.[64] The West Riding, which included Leeds and Sheffield, had seen 48 incidents in the second half of 1889 and their public impact was illustrated by reports that a crowd of 500–600 had gathered in Barnsley around a 'mad dog'.[65]

Nationally, Chaplin was said to have been naïve to apply a measure that might be justified and tolerated in towns to the countryside. Sheep farmers complained that they were unable to move their flocks with muzzled dogs and one 'Countryman' thought the measures would worsen the agricultural depression.[66] Chaplin's position was not helped by the fact that the Order did not apply 'to packs of hounds, harriers, or beagles, or to greyhounds or other sporting dogs, while being used for sporting purposes'.[67] Also, while he could claim to have steered the middle ground, the fact that muzzling remained in force in certain areas seemed to symbolise victory of the SPH. He had a faithful ally in George Fleming, who argued that the government's version of muzzling orders was the best policy and ought to be supported by all reasonable people.[68]

In fact, the muzzles did not remain on in Kent or the rest of the Home Counties for very long. The Board of Agriculture changed its Order in June 1890 from compulsory muzzling to the requirement that all dogs wear a collar and be led. Local opponents of the measures claimed that it was their petitioning and threats to vote against the government that won the day, the Board claimed that it was due to changed circumstances. The report of the Veterinary Department for 1890 set out what it termed 'highly satisfactory' returns, which showed the total incidence of rabies to have fallen from 340 in 1889 to 134 a year later, with incidents in Kent down from 34 to 2.[69] The Chief Veterinary Officer claimed that the reduction was due principally to muzzling, but acknowledged the value of improved awareness and care by dog owners, and the role of the muzzle in allowing the better identification of the 'the ownerless, useless, mongrel, stray cur'. The national figures for 1891 and 1892 were even better, falling successively to 81 and 40, with no cases in London, where muzzling orders had remained in force. The goal of the Board, the so-called 'official plan', was to reduce rabies to a minimum, not to achieve eradication.[70] The 'plan' was to use muzzles to suppress rabies and then use a registration system as a form of surveillance on dogs and their owners. It was considered politically acceptable to the public and would, if successful, eventually need the introduction of quarantine measures to keep dangerous dogs out of the country. In November 1892, the Board of Agriculture, now under a new Liberal government, revoked all Orders and handed the control of rabies back to local authorities.

Between 1892 and 1893 recorded cases of rabies in England rose slightly from 38 to 46; although the national total was higher because of two cases in Scotland, which had been rabies-free between 1888 and 1891, and had only seen three cases in 1892. However, in 1894 the returns for England increased to 229, almost entirely due to an outbreak in Cheshire, Lancashire, and the West Riding in the autumn.[71] The following year rabies reached epizootic proportions again when 672 dogs were returned as rabid. The Annual Report of the Board of Agriculture identified stray dogs as the main cause, accounting for forty per cent of cases. The 1895 outbreak was concentrated in Lancashire and the West Riding in the early months of the year, and moved to London and the Home Counties towards the end of the year. An indication of the scale of the outbreak was that between 1894 and 1895, the number of police prosecutions for unmuzzled or uncontrolled dogs rose 340 per cent to 31,434, with 27,522 convictions.[72] This data shows that local authorities and police forces did act and, whether due to muzzling or other factors, the outbreak was clearly in decline at the end of the year. Indeed, the Public Control Committee of the LCC debated whether to reintroduce muzzling in November 1895 and decided to delay as the incidence of rabies was by then sporadic. However, reports increased markedly thereafter and Muzzling Orders were re-introduced in February 1896. This decision led to vigorous actions by the police and produced a public panic. On the third day of the Order alone, nearly 800 dogs were received by the Dogs' Home. Over the whole year, the police brought in 41,395 dogs, double the previous year's total; only 19 were diagnosed with rabies.[73] The measures worked; the number of reported rabid dogs in London, Surrey, and Middlesex fell in each quarter of the year, from 129 in the first to 21 in the last.[74] The Veterinary Department attributed this wholly to muzzling.[75]

The reintroduction of muzzling orders reopened the controversies about the merits of the alternative system of collars and registration, and whether measures ought to be universal rather than the current 'imbecile system'.[76] The usual suspects from the SPH, DOPA, and the NCDL vented their familiar views; however, new opinions were expressed by kennel clubs and in the columns of the developing dog press. Kennel club members, while disliking muzzling, called for a universal 12-month order that would remove rabies and controversy once and for all.[77] In deploring the opposition to controls, an editorial in the *Ladies Kennel Journal*, stated that there are, 'too many "dog maniacs" who refuse to believe that dogs go mad' and argued that the country needed 'dog lovers not maniacs – enthusiasts but not fanatics'.[78] Nonetheless many dog lovers condemned the actions of the 'muzzle maniacs'. The NCDL led the opposition, and were later joined by *ad hoc* efforts, including an Anti-Muzzle Fund started by members of the Ladies Kennel Association, and there was a short-lived Anti-Muzzle Association.[79]

Walter Long and the eradication of rabies

In order to try to defuse the situation, in April 1896 the new President of the Board of Agriculture, Walter Long, appointed a Departmental Committee 'to inquire into and report upon the working of the laws relating to dogs'.[80] Long's background and politics were similar to those of Henry Chaplin; he came from a landed family and enjoyed country pursuits, including foxhunting.[81] His 15,000 acre estate was in Wiltshire, and he first entered Parliament as MP for North Wiltshire in 1880. He lost that and an adjacent seat in elections in 1885 and 1892, but was returned for West Derby in Liverpool in the Conservative victory of 1895. According to Alvin Jackson, in his new cabinet position 'he consolidated his reputation as a politically courageous and pragmatic reformer who was informed, but not unduly constrained by his rural constituency' and 'rabies was the issue that dominated his time at the Board'.[82] In his memoirs Long recollected that, encouraged by the Head of the Animals Division, Major Tennant, he felt able to take vigorous measures against rabies because the large Conservative majority meant that they would be in power for the five years necessary to eradicate rabies.[83] (Figure 5.3.)

The Departmental Committee, chaired by Charles Whitmore, MP, met over the summer and interviewed 27 witnesses, which included familiar protagonists like Victor Horsley, John Colam, and John Woodruffe Hill, along with civil servants, chief constables, and officials from Ireland. It

Figure 5.3   Walter Long

Source: W. Long, *Memories: By the Right Honourable Viscount Long of Wraxall, F.R.S. (Walter Long)*, London: Hutchinson, 1923, Frontispeice.

reported in April 1897 and recommended that the Board of Agriculture should adopt measures to eradicate rabies from the country and then to keep it out. In its analysis of events since the last major outbreak in the 1880s, the Committee endorsed the claim that rabies had been brought under control by muzzling and other measures when implemented by the Board of Agriculture, but had re-emerged after controls had been passed back to local authorities in 1892. Their main evidence was trends in the returns to the Veterinary Department and the comments of its officials, but they also cited evidence of lax local authorities and weak magistrates who discredited rabies controls and encouraged public hostility. Hence, a principal recommendation was that the Board of Agriculture should assume central responsibility for rabies control policy and implementation. However, the Committee did not recommend universal muzzling for six or 12 months. They maintained that such a policy was unnecessary, because many parts of the country were, and always had been rabies-free, and because it would not be accepted by the public. In their evidence, officials at the Board of Agriculture had complained about public indifference and apathy towards the threat of rabies and of police reluctance to deal with strays and dangerous dogs, because of personal dangers and public abuse. It seemed to officials that a 'rabies scare' was necessary to alert the public and galvanise local actions, which was, of course, a poor basis of concerted action – only central government could take long-term strategic decisions and see them through.

The alternative to universal measures proposed by the Department was muzzling orders over 'considerable areas', defined not by administrative boundaries, but by the likely 'marching' range of a rabid dog – a 30 mile diameter circle became the accepted range. It was also suggested that controls should not have to be continually renewed, but would be indefinite and kept in place for however long it took to suppress the disease. This was essentially the policy that the Board had tried to pursue under Chaplin and which offered a third-way between the 'maniacs' on both sides. The Committee then, seemingly confident about eradication, offered advice on post-eradication policy, advocating licensing to keep down the number of strays, the use of collars and badges, quarantines to control imported dogs, and special measures with regard to sheep-worrying and the situation in Ireland. The Committee had heard from a large number of witnesses from Ireland, where it was reported that rabies was widespread and where it was difficult to control because of the sheer number and ineffectiveness of local authorities in rural areas.

Even before the Departmental Committee had reported Walter Long had decided to act and let his intentions be known to Parliament.[84] The muzzling orders in London expired on 1 February 1897 and a report in the *Daily Mail* fancied that London dogs, 'appeared to realise the great fact of their freedom. They trotted along with erected tails, pride in their port and defiance in their eye, and to purposely go more slowly on passing a policeman, while a smirk of triumph illuminated their countenance'.[85] Any freedom was short-lived, for on 6 April the Board of Agriculture issued a new Order for London which incensed dog loving opinion, not least as there appeared to be little or no rabies in or around the capital.[86] The Order was extended in subsequent months to other rabies hot-spots in the Home Counties, Lancashire, the West Riding, and the north Midlands. There were the familiar objections on both sides. While the 'muzzle maniacs' had not been given their universal system, Long had switched government policy from suppression to eradication.[87] He recollected that it was hard to excite the public over rabies because its incidence was low, but enjoyed luck when two fearsome cases coincided with the eradication campaign.[88] However, he remained surprised by the vehemence of the opposition.

The 'dog maniacs' poured 'virulent abuse' on Long and his officials, showing greater anger than previously, but using familiar arguments about muzzles, curs, dog owners, the police, and local authorities.[89] However, five new issues inflamed opponents: the stipulation of a particular type of muzzle; the exceptions for sporting dogs; the situation of ladies' dogs; the evidence on the incidence of the disease; and Long's autocratic actions. The new Order required that dogs be muzzled with a wire cage and banned the use of the leather strap. Critics made objections at two levels. First, that the state did not have the right to issue edicts on how an owner should meet the regulations; and second, that wire muzzles were cruel and could cut the tongue and snout of a dog to pieces.[90] Many dog lovers sung the praises of leather straps, extolling their virtues as natural, flexible and adjustable.[91] The question of finding a suitable muzzle for all the different shaped snouts became a serious matter as the speed of Long's action had given no time for makers to respond. *Animal World* carried a special article, which gave details of suppliers and their specialities. However, for many dog owners wire cage muzzle seemed to combine prison and torture. One called it an 'offensive appliance' that could 'terrify and madden dogs', it was also known as 'Mr Wicked Long's Infernal Machine'.[92] 'Puzzled and Muzzled' wrote to the *Standard* to complain that his Sussex spaniel left for his walk 'a mild tractable dog' but returned 'an infuriated animal, having twisted the sides of his muzzle for that the wire pressed against his teeth and gums'.[93] Supporters of the measures, particularly Everett Millais who distanced himself from fellow members of the RSPCA, maintained that wire muzzles were highly adaptable, safe and readily accepted by dogs.[94] (Figure 5.4.)

Long's Order was described as 'class legislation' because it exempted the sporting dogs of the rich, like himself, and targeted the dogs of the poor and working dogs on farms.[95] The magazine *Our Dogs* complained that the legislation deprived the working class of many privileges and

Figure 5.4 'What in the world have you got on your heads?'
Source: *Animal World*, March 1896, 37. Courtesy of the RSPCA.

warned that 'class legislation is dangerous'.[96] In rural areas there was consternation that working dogs had no exemptions, hence, while the dogs of shepherds and farmers had to be muzzled, the hounds of landowners and country gentlemen did not.[97] However, in some instances foxhounds were affected and hunting was stopped, which Long believed showed that he was even-handed; indeed, he boasted that peers and the wives of MPs were prosecuted.[98] On the accusation of 'class legislation', Long answered emotively, claiming the higher purpose of child protection and the horrors of hydrophobia,

> never was any legislation less worthy to be so described. The comparatively wealthy could protect themselves from this disease. The people who could not were the poor, who were working, many of them, every day in the streets, and above all, the poor children, whose only playground was the streets, where they ran innumerable risks from horses and vans and accidents of all kinds, and where they ought not in the name of humanity and in a Christian country to be exposed to the appalling risk of being bitten by a mad dog, carrying with it more merely terrible suffering, but inevitable death from an incurable disease.[99]

Middle class dog owners also worried about the education and motives of the policemen who enforced the Orders, and suspected that many saw sport in confiscating the dogs of their betters. Also, as part of any fine went to their Superannuation Fund, policemen were unlikely to show owners any mercy.

Accusations of gender bias were also articulated. First, dog sports were almost exclusively the preserve of men; hence, while their leisure pursuits were unaffected, women's freedom to walk their dog had been curtailed.[100] Second, women owning small dogs with flat faces, for example pugs or King Charles Spaniels, found it particularly hard to find suitable muzzles.[101] (Figure 5.5.) Some women took their dogs out in bags and some were even fined for carrying their unmuzzled dogs in the street. One writer to *Our Dogs* suggested that Toy Dogs should be exempt as their tiny teeth would be wiped clean of virus if they bit through clothing.[102] As an instrument of physical restraint, the muzzle symbolised the difficulty that women had in having their voice heard in the public domain, as well as being another area of life in which they were subject to restraint and control. The links between muzzling and patriarchy were often made explicit. An article called 'Muzzles for Ladies' in the *Standard Magazine* in 1897 revealed the 'Gossip's bridle', 'a curious and cruel instrument of torture' employed, it was alleged, earlier in the nineteenth century by physicians and magistrates 'for the purpose of curing women of an ailment of the tongue to which they were said to be subject'.[103] The instrument gagged and locked upon the tongue, causing lacerations if the wearer dared to speak, and the 'poor creature' was paraded through the streets and became an object of scorn and derision, someone for locals to insult and degrade. Implicitly, the muzzling of dogs inherited all of these negative associations, notably the subjugation of the 'innocent' and public displays. But supporters of muzzling used the protests of women, who were said to overindulge their pets, to mock and undermine them, drawing on the trope of the over-loyal and excessively friendly lap dog and its over sentimental female owner.

A fourth new area of criticism was over the actual incidence of rabies and became quite technical. Like their fellow travellers in the antivaccination and antivivisection movements, anti-muzzlers routinely and cleverly used evidence from scientific papers against officials and scientists. As we mentioned above, when Long introduced his Order in April 1897, the experience on the streets and even in veterinary clinics was that there was very little rabies.[104] The accuracy of the returns to the Veterinary Department had been contested since first reported for 1887, and reflected uncertainties over the whole century about which dogs had been truly rabid. Anti-muzzlers suspected that rabies was over-reported in the official returns, but Veterinary Department and Long dismissed such claims.[105] In the much anticipated speech in his West Derby constituency in late April 1897, the first public defence of his policy, Long argued that the trends in the incidence of rabies, up or down, since the late 1880s had been linked to muzzling. He also pointed to the fact that veterinary inspectors followed up all incidents and undertook autopsies, and against claims that the current outbreak was imaginary, he stated that 15 cases had been verified in London already that year.[106]

The attack of official returns was pursued forensically by the barrister George Candy, who had been a founding member of the DOPA and Frederick Pirkis, who was still working to persuade rabid dog bite victims to take the Buisson treatment. In 1898, Pirkis moved the objection away from simply over-reporting to a sophisticated analysis of the Veterinary Department's case that muzzling had worked.[107] Both Long and the Department had used the falling returns between 1890 and 1892 to illustrate the benefits of central control, and blamed the rising trend in 1893–1895 on local authorities relaxing regulations. Pirkis first pointed out that only 1890 had been a muzzled year, whereas the main control in 1891 and 1892 was the collar and the removal of strays. Next, he argued that six or seven years was not enough time to identify trends in epizootics, which waxed and waned due to many factors. Finally, he pointed to the fact that over time returns had been based on three different methods. Initially, the Department had recorded reports, which might have come from any source. Pirkis wrote disparagingly that they came from, 'Well, just anyhow. Sometimes upon the word of a policeman or upon that of an ignorant very local vet'.[108] To improve accuracy, in the mid-1890s, all reports were followed up by veterinary inspectors who, in doubtful cases, used autopsies to confirm diagnoses. Then in 1897, the Department added laboratory inoculation tests in cases where autopsies were inconclusive. Cases followed up by autopsy and inoculation showed that the great majority of suspect dogs were not suffering from rabies. Pirkis argued two points; first, that the introduction of improvements was an admission that prior methods had been inaccurate and had inflated returns; and second, that the returns across the three methods used over the 1890s were incommensurable and that no conclusions could be taken from the data.

The final new challenge that Long faced was that he was being autocratic and tyrannical; indeed, his Orders were alleged to be the 'most despotic piece of legislation that has darkened the pages of

The Muzzling Order?

We muzzled our terrier pup,
But he flew at once in a rage.
When he found his head shut up
In a patent "Humane" cage.

He pawed
    and scratched his face,
Till his nose was chafed and sore.
He wriggled all over the place,
    Then rolled across the floor.

Said a mastiff
    passing by,
Your efforts
    are in vain.
But the pup made no reply,
And tugged away again.

Figure 5.5 'The Muzzling Order'
Source: *Animal World*, February 1898. Courtesy of the RSPCA.

English history since the days of King John'.[109] Such hyperbole was driven in part by the fact that he used Orders in Council of indefinite duration and because his decisions were not subject to Parliamentary scrutiny or approval.[110] Long was regularly questioned in Parliament, but he stuck to his guns on the effectiveness of the muzzle, claiming that he was following the scientific advice of his officials. An item in the *Standard* stated that the Board of Agriculture was dominated by a 'clique of experts', who 'with their eyes fixed upon the experimental laboratory, they see nothing but their own theory, which rest upon unverified, and therefore, unscientific assumptions, nor are they above juggling with evidence in order to support it.'[111] The Board's chief veterinary advisor John Tennant was subject to as many insults as his Minister, though Long was keen to remind critics that he made the decisions, not his officials.[112] An editorial in the *Times* welcomed assurances that policy was being decided by 'the logic of facts and by the learning of experts'.[113] However, it was exactly this reliance on science and the seeming absence of humanity that worried critics. For example, the *Ladies Kennel Gazette*, which had accepted muzzling in 1896, changed sides and characterised the 1897 Order as,

> A tyranny – resourceful, vexatious, pettifogging and unmanly – which is visibly creeping and growing in the body politic, which emasculates the public spirit and which is a far great danger than any physical malady. To inoculate a nation with cowardice, terror and fear is to injure it more than any bubonic plague which could be imported into it.[114]

One correspondent to *Our Dogs* argued that Britons should 'drop all pretence to humanity as nation, and go back to the old savage sports of bull baiting and bear gardens'.[115] In the popular press, Long was subjected to 'abuse' and 'ignorant clamour', which he allegedly met with 'magnificent disdain'.[116]

Walter Long seemed to relish the opposition he faced over his rabies control policy. He knew that he enjoyed the support of leading veterinarians and scientists, and the Prime Minister, Lord Salisbury.[117] In a speech in Bournemouth in November 1897, he remarked ironically on the protesters, saying that 'if he had the time he could have made by this time a scrapbook of letters, which alike for the style of calligraphy and for elegant phraseology, and for their force of language would, he believed, had proved not simply interesting but a unique collection.'[118] At the end of the year he was able to claim success for his policy.

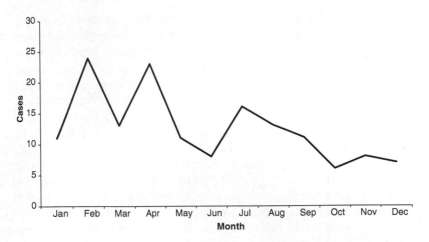

Graph 5.2   Cases of rabies in dogs each month in Great Britain, 1897

Source: Annual Reports of Proceedings Under Contagious Diseases (Animals) Acts, and Markets and Fairs (Weighing of Cattle) Acts, 1897, 1898 [C.8796] xx: 81–2 and 84–109.

Hydrophobia deaths were down from 20 in 1895 to 8 the following year and only totalled 6 in 1897. Reported cases of rabid dogs were down too (Graph 5.2) and as early as July 1897, Long was considering introducing a new Dogs Regulation Bill for the post-rabies era.[119] In September, further preparations were put in place when the Board introduced importation regulations requiring six months quarantine. These required that all dogs were imported under licence and that they 'be detained on the owner's premises and virtually isolated for a considerable period after landing'.[120] The feeling in government was that internal security from rabies was being won and that the new challenge was the external threat, with the most immediate question being Ireland.

The final push: Ireland and Wales

In January 1896 the *Times* had reported that rabies in Ireland was 'very rife' and it was suspected that dogs from there were responsible for some of the outbreaks in Lancashire.[121] Data presented to the Departmental Committee in 1896 showed its incidence had doubled since the late 1880s. Legislation in Ireland was similar to that in Britain, with implementation left to local authorities; though the fact that there were as many as 800 'districts' made concerted controls impossible. There were two problems: urban strays in cities like Dublin and Belfast, where measures similar to those in London were recommended, and wild dogs, not necessarily rabid, in the countryside that attacked sheep and game. In fact, in Ireland rabies was approached and reported as a livestock problem in officials statistics. In 1897, returns showed 497 animals 'attacked' and 1,137 'destroyed as suspected, or having been exposed to infection'.[122] Annually 3,000 sheep were lost to dogs and the effects on game were already being considered by a Committee under Lord de Vesci, who was hoping to develop the West as a tourist area, believing that if game could be nurtured it would become 'one of the great playgrounds of Europe'.[123] The threat from Ireland was said by Everett Millais to be greater than from 'the whole of the rest of the world put together'.[124] Unsurprisingly then, the Departmental Committee made the same recommendations for Ireland as for the rest of the country and set the target of eradication. This was pursued and rabies was eradicated from Ireland in 1901. However, in 1897, the country was immediately affected by the import restrictions into England and the Board of Agriculture was petitioned over the threat to dog shows and breeders on both sides of the Irish Sea.[125] Such measures were described as 'Mr Long's latest act of official vandalism' and the Irish Kennel Association immediately sought exemptions. In the event these were granted, but only on the terms that any dog imported had to be kept isolated throughout their stay.

While Ireland was the main focus of concern, the government started to worry about the near continent. There were around 2,000 cases of rabies reported in France each year and there was no prospect of eradication because of its long land borders. The continuation of imported outbreaks in the German states that had adopted stringent veterinary policing was regarded as a salutary lesson of the importance of keeping out after stamping out. The Board of Agriculture approached controlling the importation of rabies in the same way as they approached the cattle plague and foot-and-mouth, through the use of inspections, isolation, and quarantines.

The claim that rabies had been eradicated was first made in a letter to the *Times* on 31 August 1899.[126] The writer assumed that Walter Long had said as much in a Commons answer the previous day; in fact, all he had said was there had been only one case so far that year, a statement that was nonetheless greeted with cheers.[127] This success was grudgingly acknowledged in the magazine *Our Dogs*, which had been one of Long's most implacable opponents, and was celebrated by the medical profession, because 1899 was the first year for half a century with no recorded hydrophobia deaths.[128] However, hopes of eradication were premature because during the summer there was an outbreak in South Wales. Ironically, seven cases were confirmed on the very day of Long's Commons statement. Muzzling orders were issued on 8 September and the whole episode was constructed as an imported case spread by a single dog on the 'march'.[129] The Diseases of Animals Act Annual Report for 1899 noted that the outbreak was over by November and stated that 'there are strong grounds for the belief that all other parts of Great Britain have been, for a considerable period, entirely freed from the disease'.[130] In fact, the Department had received 197 reports, of which only 10 were accepted as rabies (6 on the judgement of a veterinary inspector and 4 confirmed by laboratory tests); 8 of the 10 were from the South Wales incident. In 1900 no rabies was reported in England or Scotland, but in the area around Brecon there were confirmed deaths amongst dogs, cows, a cat, a heifer, a bull, a sow, and a mare.[131] The *Times* reflected on the incident in terms of imported disease and the toll that one foreign case could wreak, and given the presence of rabies in all neighbouring countries, warned that Britain was 'surrounded by danger on every side'.[132]

Rabies in South Wales proved hard to suppress. The disease emerged again in the Brecon Beacons in March 1901 and was contained; but another case was reported in Pembrokeshire in December.[133] This outbreak persisted throughout 1902. Muzzling orders were introduced and when these failed more stringent measures were introduced, principally a night curfew and the requirement that dogs be kept muzzled indoors. The persistence of the disease was blamed on farmers who let their dogs out at night to scavenge and to the difficulties of veterinary policing in rural areas. The final cases in this outbreak were near Llandovery on 15–16 November, which turned out to be the last indigenous case in Britain until a exactly a century later in November 2002, when a naturalist died in the Scottish Highlands after being bitten by a bat.[134]

Eradication was celebrated in the annual reports of the Veterinary Department throughout the 1900s. While rabies might have been 'officially eradicated', mad dog incidents and the fear of rabies remained on the street. For example, in 1901 there were 50 suspect cases in London and 12 in Liverpool, Britain's leading centres for 'vagrant dogs'.[135] In 1900–1904 hundreds of reports of suspected rabies were received by the Department, all were investigated by inspectors using autopsies or inoculation tests, and all, except those in South Wales in 1902, proved negative.

Hounds, Holmes and hydrophobia

It was against the background of assumed eradication and alleged continuing reports of mad dogs on the streets, that Arthur Conan Doyle penned his famous Sherlock Holmes story *The Hound of the Baskervilles* in 1901–1902. He set the story in 1889, which as a medical man he would have remembered as a crisis year for rabies and the year after the Jack the Ripper murders, when it had been suggested that dog controls had made the streets less safe. Doyle may also have been influenced by the work of Catherine L. Pirkis, a fellow detective novelist.[136] Like her husband, Frederick, she was an active anti-muzzler and antivivisectionist. She had created Britain's first female fictional detective, Loveday Brooke and published in the *Ludgate Monthly*, a rival to the *Strand* in the 1890s. Interestingly, a recent review of her novels noted that she regularly used 'characters mistreating dogs to signify their brutality and/or murderous instincts'.[137] Doyle had a different pedigree, he had been on the scientific wing of the medical profession and while there is no evidence he visited Pasteur in Paris, he did report on the world's second medical breakthrough, Robert Koch's Tuberculin treatment for tuberculosis that was announced in 1891.

The Hound of the Baskervilles has largely been interpreted in terms of its gothic setting – a hellhound on a retributive mission, a tyrant aristocrat with no care for others, a family fortune, a family curse, a superstitious populace all set against the primitive 'prehistoric' landscape of Dartmoor – and references to legends such as those of Barghest and Black Shuck.[138] However, many features of the story can be read through the lens of contemporary fears about rabies, where the rational, analytical Holmes is confronted with the irrational, with myths about hounds, hysterical symptoms, and a savage beast. The events on the moor were retold by a concerned doctor, Dr James Mortimer, who travelled to London to enrol Holmes's help in solving the mysterious death of his patient Sir Charles Baskerville who had been found dead on the moors. Mortimer had discovered him with 'incredible facial expressions' and at first 'refused at first to recognise his friend before him', not because of his wounds but his tortured expression.[139] Indeed, his death was not from his injuries – as believers of the curse might have it – but from the phobia of dogs that had long troubled the minds of the Baskerville family.[140] When Holmes eventually summed up the case, he explained that the Baronet's nervous system was continually and perpetually weakened and undermined; he had been so obsessed about the curse and so infatuated with a dreadful encounter with the vicious hound, that any encounter would have frightened him to death.

Most of the story revolves around Sir Henry Baskerville, who lived in America, and was the next in line to inherit the estate. Unlike Sir Charles before him, Sir Henry Baskerville doubted the curse, but once at the Hall and living near the moor his resistance was broken. Doyle uses several tropes to show his change of mind. He heard the 'unnatural' sounds of the moor – incessant cries, howling, moans and groans – which Watson prosaically identified as the cry of birds. Sir Henry Baskerville rejected Watson's explanation as patronising, arguing that he could handle the curse. After the inevitable confrontation with the hound, Sir Henry Baskerville collapsed with exhaustion and shock, but survived. However, it was not just the family that was vulnerable. An escaped convict, Selden, was found dead on the moor, but in him Holmes identified a different mind:

> to hear a hound upon the moor would not work a hard man like this convict into such a paroxysm of terror that he would risk recapture by screaming wildly for help. By his cries he must have run a long way after he knew the animal was on his track.[141]

Throughout the story the hound was largely invisible, but retained a menacing presence. Its howling was presented as the very embodiment of raw nature on the moor: 'it came with the wind through the silence of the night, a long, deep mutter, then a rising howl, and then the sad moan in which it died away. Again and again it sounded, the whole air throbbing with it, strident, wild and menacing'.[142]

At the end of the tale, the hound leapt out of the fog to launch an attack on Sir Henry Baskerville. The hound itself was described as a monstrous sight. Watson was said to have froze and then gazed as

> Fire burst from its open mouth, its eyes glowed with a smouldering glare, its muzzle and hackles and dewlap were outlined in flickering flame. Never in the delirious dream of a disordered brain could anything more savage, more appalling, more hellish, be conceived than that dark form and savage face which broke upon us out of the wall of fog.[143]

Holmes shot the beast, but even in the stillness of death, the dog's corpse seemed animated with horror, 'the huge jaws seemed to be dripping with a bluish flame and the small, deep-set, cruel eyes were ringed with fire'.[144]

The mystery dissipated when Holmes revealed the criminal behind the curse, Stapleton, the Baskerville's house servant. He had frightened Sir Charles to death and intended a similar fate for other Baskervilles. Indeed, it emerged that Stapleton was a long lost cousin and also a Baskerville. Holmes, in his summation to Watson, described the hound and the curse as 'a cunning device for driving your victim to his death'. Doyle wrote within and outside the Gothic tradition. In keeping with the former, he set his story in a less civilised place and in typical fashion the curse was an echo of the past intruding upon the present, with its tyrannies, legacies, and unwelcome survivals or returns. However, Holmes's account of events was also modern and materialistic. He explained the curse as a mental disorder, both in relation to the mental weaknesses and phobias in the family and the specific, degenerative, criminal pathology of Stapleton. We would also emphasise Doyle's use of key contemporary features of hydrophobia – the contorted face in death, possible hysterical symptoms, and of rabies – the haunting howl, the dog on the march, flowing saliva and the 'muzzle', a word which referred to both the dog's snout and the technology of restrain. Twice in the book, Watson reports feeling a sensation of choking in his throat, and reported acute sensitivity to atmospheric changes. As we have shown, these were often reported as symptoms of hydrophobia, and Doyle was invoking a sense of claustrophobia and mental fear in the book, by stressing these notorious hydrophobia symptoms to describe Watson's anxieties and changes in mood. Doyle's readers, given their recent and current experiences of 'mad dogs', mad dog panics, the struggle over muzzling, and contemporary fears about hydrophobia, would perhaps have found *The Hound of Baskervilles* a richer and more resonant story than modern commentator and readers have realised.

Conclusion

Members of the NCDL were said to have 'rejoiced' when muzzling restrictions were lifted from the metropolitan area in November 1899. They would also have been pleased with the announcement of eradication, though they doubted that government measures were responsible and continued to bait Victor Horsley and the SPH. However, politicians, officials, and scientists claimed the credit. Walter Long in particular became something of a celebrity and enjoyed the plaudits for his campaign. At one extreme, the Veterinary Department spun the official line, since largely accepted by historians, that eradication was due to the enforcement of muzzling. At the other extreme, anti-muzzlers like Frederick Pirkis implied that the incidence of rabies had been 'reduced' by changes in the basis of the official returns, as these switched from police reports, to autopsies by veterinarians, and finally to inoculation tests. Each new test was more stringent, reducing the number of 'accepted' cases and confirming his view that rabies had always been rare.[145] The favourite question of the anti-muzzlers was posed again: why, if London police-men dealt with so many rabid dogs, had none ever developed hydrophobia?[146] Indeed, the Commissioner of the City of London Police who had been responsible for dog controls, Sir Henry Smith, wrote in *Blackwood's Magazine* in 1900, that he was an 'Unbeliever'.[147] He seemed to doubt that rabies had ever been rife in the last 46 years when the police had handled thousands of dogs; he thought epilepsy or tetanus were often mistaken for rabies. The accusation was that the govern-ment's achievement might be far less impressive than they claimed; hence, it was possible that other changes, such as better diagnosis, the reduced number of strays, or the better care of dogs that had been the decisive factors in eradication.

The truth probably lies in a combination of factors and changes over the whole decade and not just from 1897. First, once rabies was added to the CD(A)A in 1886, the Veterinary Department increasingly approached it as they did other contagious animals diseases as some-thing to be contained, stamped out, and kept out. Retrospective accounts of eradication also ignore the fact that the muzzle was a cum-bersome piece of technology. It was of little use as a restraint on a rabid dog. Contemporaries complained about the availability of muzzles, the lack of standardisation and the ability to get a good fit for their dog's muzzle. The power of muzzling was less a direct method of preventing rabid dogs biting than a reassuring symbol of administrative control of the problem, and distinguishing animals that were disciplined and ordered from those that were not. Until 1897 the government's capacity for decisive action was constrained, but thereafter its officials deployed not only the muzzle, but also stricter surveillance, expert investigation of all reports (made easier by reduced incidence), and advanced laboratory methods. While at one level, officials were working with blanket meas-ures, like widespread muzzling and import controls, they increasingly used more precise and targeted measures, such as dog curfews and the control of movements. Second, there is evidence of changing public atti-tudes to dogs – owners and their dogs became more obedient. A higher proportion of owners bought dog licences, which the Assistant Secretary to the Chief Veterinary Officer thought was 'probably due to a growing sense on the part of owners of dogs of their responsibilities'.[148] The gov-ernment was helped by the relentless campaigns to clear the streets of stray dogs and the promotion of responsible ownership by animal welfare organisations, the new pet supply industries, and popular images of the loyal, affectionate domestic pet.

6
Rabies Excluded: Quarantines to Pet Passports, 1902–2000

Rabies re-entered Britain only twice in the twentieth century, in 1918–1922 and in 1969. However, mad dogs and their disease remained in the popular consciousness throughout the century, not least because of the national pride over being a rabies-free country and propaganda on the threat of imported rabies. This was evident at all levels of society, from government ministers, civil servants and the scientific elite, through to antivivisectionists, who claimed that it was their protests that prevented a Pasteur Institute being established in London and had produced the stamping out policy. In the inter-war period, rabies became exotic, even tropical, a disease of the dogs of expatriates return-ing from India and Asia. It was, of course, the association with Empire that Noel Coward drew upon in his parody of British colonial society in his song 'Mad Dogs and Englishmen'. Yet, in the last quarter of the cen-tury, rabies was re-domesticated in a European context, with attempts to reinforce border controls and quarantines against the new threat from fox rabies, and growing support for allowing vaccinated animals to be imported without quarantines. Indeed, the year 2000 saw the partial relaxation of the iconic quarantine regulations and the introduction of Pet Passports. These allowed dogs and cats that had been vaccinated, passed blood tests, and microchipped, to be brought in and out of the country.

Dogs of war

Walter Long dined out on his role in the eradication of rabies for much of the Edwardian era, at a time when his career became more controver-sial because of his unionist stance on Irish affairs.[1] The policy was por-trayed as an example of decisive, rational action, and of standing out against opposition to do the 'right thing' for the country. When commenting on his political style, the liberal literary weekly *The New Age* observed, 'Mr Long never did anything but muzzle. Because muzzling succeeded with hydrophobia, he thinks, like the quack statesmen he is, that muzzling will succeed with everything else.'[2] Although the country was officially rabies-free, there continued to be reports of the disease. In 1903, the first post-eradication year, there were 110 reports of rabid dogs in 21 counties; each was investigated and in suspect cases laboratory tests were made – all proved negative.[3] Reports of rabid dogs continued. In 1904 three English dog bite victims were treated in Paris and 49 allegedly rabid dogs were detained in London.[4] In 1906 after a scare in Essex, 21,380 dogs were destroyed and as late as 1914 the police in London apprehended 12 suspect dogs.[5] Rabies might have been officially eradicated, but it remained on the streets, in the courts and in the public mind.

Long's successors at the Board of Agriculture, Conservative and Liberal, came under pressure from dog owners, especially wealthy owners who travelled abroad regularly, to relax quarantines so that imported dogs could be held at home rather than with veterinary surgeons.[6] One correspondent to the *Times* in 1908 contrasted the suffering during separation of 'carefully tended and delicately nurtured pets', which 'would naturally pine and fret', with the fact that yard dogs and sporting dogs would be untroubled as they were used to kennel life.[7] In addition, the absence of any reported rabies cases amongst quarantined dogs suggested that once again it was responsible owners who were complying with the law, while illegal imports of potentially more dangerous dogs went undetected. The Board kept faith with the strict regulations, a policy that was helped by occasional reports of rabies outbreaks, as in Northampton in March 1908 and by high-profile breaches of quarantine as in 1910 when Mick, a 'Canine Houdini', became the most wanted dog in England after he escaped from kennels in Hendon.[8]

There were reminders of the horrors of the disease in reports of Britons dying overseas and of people suffering terrible deaths on their return home, as in the case of two soldiers who were bitten in Gibraltar in September 1910 and sent to Paris for treatment. One died within three weeks, but the second, George Seaman returned home and joined the city police only to die six months later of rabies in Hackney Infirmary.[9] However, concessions were won by dog owners; the most significant was to reduce the six-month isolation when an owner declared that their dog had been under 'person control' for three months prior to shipment. In practice, many owners were only too willing to sign such a declaration and returns showed 60–70 per cent of imported dogs left quarantine early. Dogs were also allowed to stay 'at home', either if their owner was 'seriously ill or that his or her health was seriously affected by separation from the dog', or if secure premises were built at home and kept under the supervision of a veterinary surgeon.[10]

At the start of the Great War Walter Long was back in charge of rabies policy as Minister of Agriculture. He was advised in October 1914 to reduce the quarantine to four months, but to apply it strictly to all dogs and only allow exceptions for dogs from Australia, New Zealand, and Jamaica, and dogs that had been on board ship.[11] The Defence of the Realm Act, the legislation that restricted civil liberties during the First World War, amongst other things allowed government officials to destroy stray dogs and banned dog shows.[12] Generally, dogs had a good War.[13] They were used on the front in a variety of roles and some were given awards for bravery. But in France control measures broke down; the number of rabid dog incidents in Paris increased from 3 in 1913 to 62 in 1916 and 411 in 1918.[14] The number of French citizens treated at the Pasteur Institute also rose, from 92 in 1914 to 350 in 1918.

Rabies returned to England in the summer of 1918. Mad dogs were reported in south Devon over the summer and the presence of the disease was officially confirmed by the Board of Agriculture in September.[15] The authorities assumed that it had been imported by soldiers returning with pet dogs they had befriended while fighting in France. The Board sent a Committee of Inquiry to investigate the outbreak in November 1918 and they concluded that quarantines had been ignored due to the large movements of troops, the congestion at ports, and the activities of Royal Air Force pilots.[16] They dismissed claims that the disease had arisen due to food rationing causing dogs to be starved or given rotten food. However, elsewhere in government such worries seem to have been taken seriously as the Interdepartmental Committee on Dog Food did arrange for maize to be released to improve the quantity and quality of dog biscuits.[17]

The first case had come to the attention of the police on 18 August, when a suspicious dog death was reported by a leading local clergyman, Dr Trelawney Ross. Rabies was confirmed by laboratory tests and Dr Ross and his family were sent to the Pasteur Institute for treatment; prayers were also said in local churches.[18] Local authorities introduced control orders and the Board placed restrictions on the movement of dogs. These measures were soon strengthened by the addition of muzzling, and there were night curfews on dogs to try and halt their movements in rural areas.[19] The familiar sights of rabies control measures were soon evident: the rounding up and destruction of strays, public complacency, dog owners summoned to court, criticisms of leniency by magistrates, and complaints from dog owners of unnecessarily harsh measures.[20] The police and magistrates were encouraged by local and national politicians to be more severe and the number of dog owners appearing in court increased, with accidental breaches dealt with as severely as deliberate flouting of the law.[21]

The outbreak moved west over the winter, which along with the fact that most animals had paralytic rabies and tended not to wander or bite, made it easier to control.[22] The country was still subject to many wartime controls, so the public was assumed to be more likely to comply with restrictions. There were, of course, exceptions and a particular problem was the evidence that wealthy lady dog owners, who were 'marooned' in their holiday homes and unable to return to London, resorted to smuggling dogs under their skirts.[23] From the start of the outbreak dog bite victims were sent to Paris for treatment, but such was the demand, together with the problems of travelling through northern France, that arrangements were made by the Local Government Board (LGB) for treatment to be available in Plymouth. This service began on 1 January 1919.[24]

Rabies in the south-west declined rapidly in prevalence and profile in the spring of 1919, and attention switched to new outbreaks in south Wales and London.[25] Treatment centres were established in Cardiff and Surrey, and anticipating the further spread of the disease, the LGB decided to produce their own vaccines. It obtained *virus fixée* from Paris and employed David Semple, former head of the Kasauli Institute in northern India, to make the vaccines.[26] Government preparations did not prevent a rabies panic over the Easter period. The London outbreak began on 14 April when four people were bitten in Byfleet, Surrey, which led the implementation of a Rabies Order and restrictions on the movement of dogs over a wide area including London, Middlesex, Surrey, and adjacent parts of surrounding counties.[27] Within days there were further reports of mad dogs in Acton, Ealing, Bermondsey, and Catford, and in the West End. Many incidents were front page news in the popular press.[28] In the West End incident, it was reported that an apparently mad dog ran up Oxford Street 'with foam dripping from its hanging tongue and its eyes rolling wildly', before darting into an apartment where it was shut in a scullery by a young girl. The mastiff then 'held four police officers at bay for four hours' before being lassoed and killed.[29] Other mad dogs were similarly dispatched and the Dogs' Home at Battersea was soon overrun as double the usual number of dogs were brought in.[30] The press gave advice on what to do if bitten and on the controls.[31] A particular problem was that the new Orders came into effect just before Easter and the holiday plans of many Londoners were disrupted as the railways refused to issue dog tickets, and cars leaving the capital were searched by the police.[32] Owners of lap dogs, reported pejoratively to be mostly women, contacted the Board of Agriculture to ask for exemptions, arguing that 'it was the height of absurdity to suppose that such inoffensive creatures could contract rabies' – but to no avail.[33]

Dog owners had great difficulty obtaining muzzles and the government came under pressure to ensure that there was adequate wire available to prevent profiteering – a common complaint in the war – and to stop owners unable to afford or obtain muzzles choosing to 'get rid of' their pet.[34] Many dog owners improvised, with varieties of muzzle lampooned in a cartoon in the *Daily Express*, which also illustrates how the outbreak was met with humour rather than horror.[35] (See Figure 6.1.)

The outbreak ended as swiftly as it began. No further cases were reported in the Home Counties after ten days, though the control orders remained in force for many months.[36] There were reports in May of rabies in Lancashire and Yorkshire, which led to calls for universal muzzling of the whole country by the Society for the Prevention of Hydrophobia and the Reform of the Dog Laws (SPH) and the Kennel Club.[37] For a time it seemed as though there would be a rerun of the controversy of the late 1890s: Frank Karslake, on behalf of the SPH, published a book advancing their cause, which was answered by Ethel Douglas Hume's *Hydrophobia and the Mad Dog Scare: An exposure of Pasteur and Pasteurism, etc.*[38] By July reports had diminished further, though muzzling orders remained in place. Over the year, 179 people were bitten by dogs in the controlled areas, of which 46 were bitten by dogs proved to be rabid.[39] No one died, which was said to be 'a triumph to the memory of Pasteur'.

The outbreak in the Home Counties coincided with an attempt by antivivisectionists to legislate against the use of dogs in laboratory experiments. The Dogs' Protection Act was introduced as a Private Member's Bill in March 1919 and was debated on 23 May.[40] It seemed as though there was to be a repeat of the discussions around the Royal Commission on Vivisection which sat from 1907 to 1912.[41] The National Anti-Vivisection Society supported the Act on moral grounds; its spokesman Stephen Coleridge stated that 'it would be far better for the world be without more physiology than more pity'.[42] Scientists again argued that they avoided painful experiments, that very few dogs were used in their laboratories, and that the benefits to humankind were

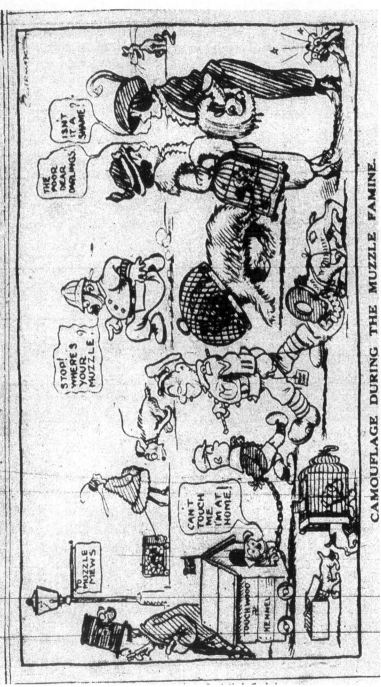

Figure 6.1 'Camouflage during the muzzle famine'

Source: Daily Express, 24 April 1919. Reproduced by permission of Express Syndication.

worth the minimal suffering endured by animals.[43] The rabies outbreak was a boon to scientists opposing the legislation; they cited Pasteur's anti-rabies treatment as the great triumph of vivisection, though they qualified this by pointing out that rabbits rather than dogs were now used in the work.[44] The connection was cited in an editorial in the *Times*, which agreed with the surgeons Sir Charles Ballance and Walter Spencer, that it was a disgrace that Britain was 'almost the only civilized nation' without a Pasteur Institute.[45] It concluded with the hope that controls on animal experiments would be eased rather than tightened. Support for vivisection could be found in the *Church Times*, where a test was set for the antivivisectionist: 'In the event of his little daughter being bitten by a mad dog would he refuse to allow Pasteur lymph to be employed? ... the consistent antivivisectionist in order to spare a few rabbits a little pain is willing to sacrifice the life of his own child.'[46] For the distinguished doctor and Belfast MP, Sir William Whitla, Britain's standing in the world and the future of science was at stake.

> Not a word was said about the thousands of human lives which had been saved as a result of inoculation for hydrophobia. ... As a result of legislation this country was already behind in the march of progress for the relief of human suffering. When he thought of the result that had followed experiments on animals he was, as an Englishman, positively ashamed of the sickly sentimentality shown in this country. Were they willing to take a back place in the noble race for the relief of human suffering and the protection of life? Were they willing to drive our best intellects to the Continent to practise in the laboratories of the Hun or alongside Bolshevists in Vienna?[47]

The government was opposed to the Bill and moved an amendment, which was passed, to change its title from to 'prevent vivisection of dogs', to 'impose further restrictions on the vivisection of dogs'. When the Bill came up for its third reading in July, Sir William Watson Cheyne, formerly Joseph Lister's closest lieutenant, senior surgeon and MP for universities of Edinburgh and St Andrews, moved that the Bill be dropped, and it was voted out after heated discussion.[48]

There were sporadic reports of rabies in Wiltshire, Dorset, Hampshire, and Surrey in 1920. The next outbreak was in 1921 when there were 14 cases in the Southampton area.[49] A further case was confirmed there in the summer of 1922, supporting the opinion that the source of outbreaks was the smuggling of dogs on ships and by private aircraft.[50] The SPH had continued their lobbying and argued in 1921 for imprisonment rather than fines to deter smugglers.[51] Press reports concentrated on those who should have known better and the lengths people would go to avoid quarantines. For example, a missionary doctor, on leave from China, claimed in his defence that his Pekingese had never mixed with other dogs.[52] In court he admitted to considering hiding the dog in a collapsible basket or in the pocket of his Burberry, but in the end he acquired a rope ladder which he used for a clandestine escape from the ship. Other ingenious methods were employed. In his autobiography, Sir Frederick Hobday, who had been principal of the Royal Veterinary College(RVC), told of owners who used morphia to sedate pets they carried in bags or pockets, and of women making use of 'a portly bosom' or hiding their 'stoutness' under a loose cloak.[53] He wrote that customs knew of a case where 'A dog "coming to" too soon, secreted in a lady's blouse, had given the game away by suddenly barking or poking its nose to have a look at the world outside'. To make his point about the power of sentimentality, Hobday also told the personal story of a wealthy lady who had promised a donation of £50,000 to endow the Canine Chair at the RVC. She withdrew the offer in her last days in a codicil to her Will, because he had once refused to let her dog out of quarantine early. Hobday commented, 'The authorities will make no exceptions whether the owner be pauper or prince ... a lady in her position ought to have recognized it, but the sequel was disastrous for College finances'.[54]

Animal welfare groups continued to campaign against quarantines arguing that: they ought not to apply to imports from rabies-free countries; they were too long; and respectable owners ought to be trusted to keep their dogs isolated at home.[55] Quarantines had been increased from four to six months in 1918, and the demand was so high that isolation was now provided by private kennels as veterinary practitioners had been unable to cope. Pressure for relaxation in the rules for specific groups continued to be made and critics pointed to the fact that between 1922 and 1927 only four dogs died in quarantine; all were terriers, they came from India and Egypt, and died between 3 days and 5 ¾ months of entering quarantine.[56] In 1927, a group of ladies lobbied for a relaxation in the length of what they pejoratively termed the 'solitary confinement' suffered by dogs to three months, though they were keen to stress that they appreciated the security from rabies enjoyed by their dogs.[57] The Minister of Agriculture, Walter Guinness, answered immediately quoting instances of incubation periods of over three months and, while recognising the problems caused to owners and dealers, rejected any change in regulations.[58] Everyone knew that quarantines were commonly breached, so some veterinarians and doctors began to contemplate a different policy, the protective vaccination of dogs, following precedents in Japan, Italy, and California.[59] However, this went against the established national policy of controlling animal diseases by stamping out and keeping out.[60] Britain's rabies-free status was always cited by governments as an example of an unpopular measure that had seen through to bring long-term benefits which everyone acknowledged. It was also emblematic of Britain's island situation, and of the country's ability to eliminate foreign plagues and to defend itself against invasion.

In the 1920s and 1930s, rabies and hydrophobia became increasingly exotic diseases both in time and space. Domestically, they belonged to earlier times of weak government, unsophisticated science, unregulated streets, and, most recently, to the abnormal conditions of war. At the first international congress on rabies organised by the League of Nations in Paris in 1927, the very different situations in other countries was apparent.[61] The most striking was the Soviet Union, where it was reported that in 1925 alone over 23,953 people had been treated in the country's anti-rabies institutes.[62] This was reported as a ten-fold increase over a decade and an implicit criticism of the communist regime. As ever, there was a fascination with the role of wolves and the prospect of a state campaign to reduce their numbers. In the inter-war years there was a concerted effort by several Continental governments to control rabies through measures that targeted street dogs and irresponsible owners. These measures were largely successful, the last human death from a rabid dog bite in France was in 1924, and by the end of the 1930s rabid dogs were quite rare on the near Continent.[63]

That expertise on rabies in humans and animals lay in the Empire had been evident in the 1918–1922 outbreak when the government called on Sir David Semple to advise on vaccines.[64] At Kasauli he had developed new methods to prepare and store the vaccine in ways appropriate to Indian conditions, that were set out in the Local Government Board's Report in 1921.[65] India remained the main centre of rabies research in the Empire, with a second Pasteur Institute established at Coonor to serve Southern India. These institutions treated thousands of dog bite victims every year, studied the efficacy of different vaccination regimes and supplied vaccine to hospitals.[66] In the 1920s, rabies emerged as a problem in African, Far Eastern and West Indian colonies, with particular interest in events in Trinidad where a mysterious new disease had been causing alarm since 1929.[67] In 1931 bacteriologists in the colony confirmed the disease was rabies and that it had been spread by bats – the first recognition of what is now recognised as a major reservoir of the disease.[68] The story was taken up in the press with the *Times* reporting on a 'strange' and 'mysterious' disease, and such was the spectre that the *New York Times* carried the story under the headline 'Mad Vampire Bats Spread Hydrophobia'.[69]

Scientific and medical work on rabies did not flourish in the inter-war period. One problem was that the rabies germ was an 'invisible', 'ultra-microscopic', or 'filterable' virus. It could not be seen with a light microscope and could not be cultured, like bacteria. The modern idea of a 'virus' as an intracellular infective agent was still in the making and there was no consensus over whether viruses were complex chemicals, bacteria-like organisms, or something else until the late 1920s, and spontaneous generation was back in play.[70] Sir Henry Dale, head of the Medical Research Council's National Institute for Medical Research at Mill Hill, London, wrote in 1931 that 'I should not myself regard the spontaneous origin of rabies as out of the question under certain circumstances [such as] extreme environmental changes, such as temperature or fasting'.[71] As well as being difficult to work with, the virus was dangerous and the low human death toll meant that it was only a public health priority in Asia and eastern Europe. However, work on the improvement of treatment and protective vaccines continued across the world and the 1927 Paris meeting heard of over 14 variations on Pasteur's vaccine.[72]

The control measures implemented in northern European countries, particularly France, ensured that the disease was not the same problem in the Second World War as it had been in the First. British quarantine regulations were very effective and while nine dogs died in quarantine between 1945 and 1949, there were no deaths between 1950 and October 1965 when a leopard imported from Nepal died after one day.[73]

It was not until the 1960s that work on rabies caught up with that on other pathogenic viruses.[74] Then the form of the virus was constructed from electron-micrographs and its pathology demonstrated.[75] The causal agent – a Rhabdovirus – was found to be bullet-shaped, and experimental research confirmed that it was usually transmitted in the saliva of an infected animal. Scientists established that the long and variable incubation period was due to the fact that, in order to produce the disease, the virus has to enter nerve cells, which it did directly through a bite or after replicating in other tissues. The virus then had to be transported passively and hence slowly, through nerve cells and synapses to the central nervous system. On reaching the spine and brain the virus multiplies and spreads rapidly, to produce encephalitis and the characteristic symptoms of rabies.[76] This understanding has remained stable into the twenty first century, complicated only by the recognition of different strains of the virus and the impact of molecular biology.[77]

Fritz and Whiskey: rabies returns

Rabies emerged from its low post-war profile in Britain in October 1969. In the early part of the year, the World Health Organisation (WHO) had warned of the 'particularly disquieting' epizootic of fox rabies that had started in Poland in the late 1940s and had crossed Germany and was now in eastern France.[78] Thus, Europe faced a new problem, not the dog rabies of the nineteenth- and early twentieth century, but a disease of wildlife. A report in *Nature* warned that there was a large enough reservoir of susceptible animals in Britain for a new epizootic to develop from just one imported case.[79] However, it was not German foxes, but an imported dog from Germany that put rabies back in the news and in the political arena towards the end of the year. A terrier named Fritz died of rabies at Camberley in October 1969. It precipitated something akin to an old-style rabies panic and led eventually to a major government inquiry.

Fritz was brought to England on 4 April 1969 by an army family returning from Bielefeld in West Germany.[80] He went into quarantine at Caesar's Camp kennels at Folkestone and was released on 4 October. A week after leaving quarantine he started to behave peculiarly, hiding himself under a bed and refusing to come out. When dragged out, his hind legs were paralysed and he refused food and drink. The following day Fritz became excited and aggressive, which led to a modern mad dog chase in suburban Surrey.

> On the 14th October, at about 7.45 am, Fritz escaped from the house of its owner, killed a cat owned by a neighbour and, after biting the boot of a milkman, ran off and disappeared. At 8.35 am, some 50 minutes later, Fritz was seen climbing into a taxi full of school children. The owner recovered the dog from the taxi, being bitten on the hand and leg in the process, and held it [in the bathroom] until it was removed to the house of the Army veterinary surgeon who reported the case.[81]

Fritz was confirmed as suffering from rabies the next day and preventive vaccinations were offered to the family and contacts. It turned out that he had been taken to the local primary school a few days earlier and had been stroked and handled by children, so there was great anxiety amongst parents and the health authority. Restrictions were imposed on the movement of dogs in the area and muzzling was required. While there was no evidence that Fritz had been 'on the march', the Ministry behaved as if he had, taking typically draconian measures with a cull of local wildlife through poisoning and shooting.[82] A 'mass extermination' of local wild life was organised on 30 and 31 October, which bagged 11 foxes, 102 squirrels, 1 rabbit, 5 jays, 7 magpies and a crow. (Figure 6.2.) There were no further cases in the area but the episode was taken as warning of the dangers of rabies and quarantines were extended to eight months (Figure 6.2).

Within a fortnight rabies was back in the news. On 11 November a black Labrador called Whiskey, also imported from Germany and held in the same kennels as Fritz, showed signs of the disease. He was put down two days later and tests confirmed the diagnosis. The Ministry set up an enquiry to determine whether the two cases were a coincidence, or whether there had been cross-infection at the kennels, either between Fritz and Whiskey, or between both dogs and a collie imported from India, Bob Judd, who had died of rabies at the kennels in July.[83] The report, by Dr Henderson and Mr Beynon, was inconclusive, finding no irregularities at the kennels and no evidence of contact between the dogs. Two explanations were offered, with each said to be a remote possibility: either both dogs had been independently infected in Germany, or that there had been some form of indirect transmission at the kennels.[84] However, as soon as the report was published the Ministry received allegations about poor standards at the kennels and that Bob Judd had escaped from his kennel and mixed with other dogs. A new inquiry was set up and this heard evidence that the strict separation of dogs was regularly breached and that staff were generally negligent.[85] The owners contested these claims, but the second report rejected their version of events. In January 1970 the Ministry stopped issuing licences to Caesar's Camp Kennels and its veterinary surgeon was disciplined.[86] Normally the kennels would have been closed, but there was a national shortage of quarantine places because of the extension of the quarantine period.[87] In December, the government tightened controls further, announcing a ban on the importation of exotic animals, a measure that had been asked for the Federation of Zoological Gardens.[88]

All was quiet until late February 1970, when a bitch, Sessan, died of rabies in Newmarket, an event that made the front page in the popular press.[89] She had been imported from Pakistan in May 1969, had left quarantine in November and died nearly nine months after arriving. This was seen by Ministry and veterinarians as a very worrying case as it

Anti-rabies shoot at Camberley

The first of the two shoots in the anti-rabies extermination drive at Camberley, Surrey, took place on Barossa Common yesterday. Men of the Royal Ordnance Corps acted as beaters for Ministry of Agriculture marksmen, shooting with 12-bore shotguns. Left: There were almost as many cameramen as guns as the marksmen set off. Below: Army beaters make their way across Barossa Common in the autumn sun

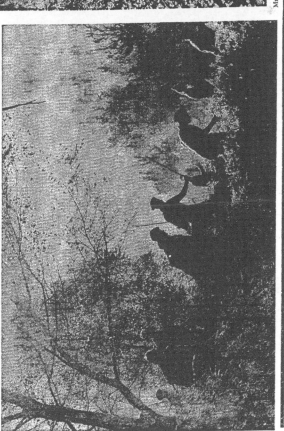

Mr. Carrington with the 20lb vixen he shot on the first drive through the common

Figure 6.2 'Anti-rabies shoot at Camberley'

Source: Times 31 October 1969, 16a–f. Reproduced by permission of News International.

would not have been caught even by the new eight month quarantine.[90] The government responded with a ban on the importation of all dogs and cats, which was in part a response to speculation that the world was facing a new rabies virus with a longer incubation period.[91] The *Daily Mirror* led on the story under the headline 'War on Pet Smugglers'.[92] The ban almost certainly originated from a recommendation of Sir Solly Zuckerman, the government's chief scientific advisor, who had written to the Prime Minister, Harold Wilson, on 4 March 1970 urging swift and decisive action. Zuckerman had been chief scientific advisor to the Ministry of Defence, before taking over his larger role in 1964.[93] He was a distinguished zoologist, who also served 'in effect as chief executive officer' of the Zoological Society of London, which ran London Zoo in Regent's Park and Whipsnade Zoo in Hertfordshire.[94] At this time he was trying to modernise both establishments, which were facing competition from new wildlife parks, which Zuckerman dismissed as commercial rather than scientific establishments. Privately, he was worried about standards of animal welfare at the new parks, most of which were run by the leading circus family, the Chipperfields, and the alarming fact that foxes were regularly seen mingling with lions, giraffes, and wildebeests.[95]

The Cabinet discussed Zuckerman's memo on 5 March and as well as confirming the import ban, set up a Committee of Inquiry chaired by Ronald Waterhouse, QC, with Zuckerman its most influential member. The Committee accepted written evidence from various national and international bodies, and interviewed experts and representatives of interest groups, including airlines, circus proprietors, fur breeders, kennel owners, laboratory animal breeders, the pet trades, scientists, and veterinarians.[96] Members visited eight quarantine kennels and other sites, including Blackpool Tower Circus, the Central Veterinary Laboratory, London Airport, Woburn Safari Park, and London Zoo. The Committee published interim recommendations in July 1970 and its final report in June 1971. The Interim Report suggested lifting the total ban on imports, moving quarantines back to six months, and requiring imported animals to be vaccinated. The government acted promptly to reduce the length of quarantines, creating what were described as scenes of 'joyful reunions' between pets and owners.[97] Both reports supported the regulations that had been in force for most of the twentieth century; that is, a six-month quarantine, rigidly enforced with 'no exemptions'. The final Waterhouse Report set out in great detail procedures for the inspection, transportation, and quarantining of different species. The major difference with the Interim Report was that the call for the vaccination of quarantined animals was dropped with no explanation.

Instead, it emphasised the need for an elaborate quarantine bureaucracy and for this to be modernised to deal with the fact that the great majority of dogs and cats, not to mention exotic species, came into the country by air.

The Committee welcomed the fact that the public had been primed for its findings by an episode of the BBC's Doomwatch series, entitled 'The Inquest', shown on 1 March 1971.[98] The programme featured the death of a ten-year-old girl and the public outcry when the destruction of all dogs in a 5-mile radius was ordered.[99] Such popular representations of the problem emphasised the point that the main threat was from dogs; however, the Waterhouse Committee's Report did cover other pets, exotic animals, and foxes. On the latter, the Committee cited data on how the situation in northern Europe had changed since the 1920s. Then, 99 per cent of reported cases in France and Germany were in dogs and domestic animals, while by the 1960s the figures were 7 per cent dogs, 15 per cent other domestic animals and 78 per cent wild animals, mostly foxes.[100] Yet, experts were divided on how easily fox rabies would transmit to pets.[101] On the one hand, some warned that 'a single case in a fox might result in an enzootic situation' and Waterhouse himself stated that fox rabies was likely to reach the Channel coast in the 1980s.[102] On the other hand, the evidence from Europe was that the fox virus was not easily transmissible to other species, in part because the fox is a scavenger and is only a predator of small animals like mice, rabbits, and birds.[103] With opinion equivocal, this aspect of rabies did figure prominently in the recommendations. The Waterhouse Report was received favourably by the government which promised to act on its recommendations, though by the summer of 1971 the 'crisis' had passed and it was only news for a day.[104] However, it was not long before foxes and rabies were back in the news, prompted by worries about foreign travel, mass tourism, and the Channel Tunnel.

Foxes and France

Britain joined the European Community (EC) on 1 January 1973. The anticipated rise in the volume of trade, travel, and tourism was expected to increase the risk of the importation of rabies.[105] All the more so because of the westward march of fox rabies was reported in the British press on maps, with shaded areas and sweeping arrows, that echoed the advance of the Nazi forces in the Second World War. Such maps were also used in the opening credits of the popular comedy show *Dad's Army* which was then on prime time television. In July 1973, the British

Veterinary Association (BVA) and conservation organisations made a concerted call for stronger measures to police the ports and to punish 'irresponsible smugglers'.[106] The usual groups were suspects: the ignorant, the sentimental, the selfish, and the rich. For those who simply could not bear to be parted from their pet, smuggling was a one-way bet monetarily, as the fine was normally less than the fee for quarantine.[107] The increased traffic on Channel ferries multiplied the opportunities for evasion, with caravan owners hiding pets behind panels seen as likely offenders.[108] Private yacht owners, who enjoyed the freedom of the seas, were suspected of believing themselves to be above the law.[109] In practice, the largest group of offenders were military personnel returning from abroad, which may simply have been because their movements in and out of the country were better monitored. In 1974 it was reported that 206 people were caught trying to import pets illegally; how many had avoided detection was, of course, unknown.[110]

The government responded to calls for tougher measures in 1974 with a new Rabies Act that created designated landing ports for imported animals, extended the species subject to quarantine (a response to the fashion for exotic pets), increased the maximum fine for breaking the law to £400, allowed offenders to be sentenced to a maximum of one year in prison, but did permit licences for some larger animals that would be housed in zoos.[111] The Act also included regulations to deal with a post-quarantine outbreak, which included controls on animal movements and powers for the destruction of wildlife in affected areas.[112] The latter measures were seen by many observers as preparation for the 'inevitable' entry of fox rabies; however, the government's line was that its policies were designed to give the country absolute protection. Ministers raised the possibility of allowing local authorities to destroy any animal imported without a licence, though this was resisted as critics pointed out that the innocent animal rather than the guilty owner would be punished.[113] Nonetheless, the Ministry of Agriculture and the veterinary profession encouraged a siege mentality, where the invasion of rabies, through fifth-column smugglers or the incursions of wily foxes, had to be resisted by measures that were demonstrably tough and might include the summary execution of pets.[114]

Rabies was back in the headlines in the summer of 1975. The WHO warned that fox rabies was advancing 20–30 miles westward each year, which was reflected in headlines such as 'The deadly virus marching across Europe'.[115] The government produced multi-language posters to be displayed at ports in Britain and Europe, with leaflets, and television and radio campaigns before and during the holiday season.[116] (Figure 6.3.)

Figure 6.3 'La Rage', RP3 (F), Central Office of Information, 1976. This poster was displayed at French ports and airports. It was also produced in English, German and Spanish.

The horrors of rabies were featured widely in the press in June, when two men died from the disease in London having been bitten abroad.[117] Their illness was linked by the press and their doctors to the advancing European 'frontier' and they hoped that graphic details, such as one victim's fears of 'drowning in this own saliva', would shock potential pet smugglers.[118] In January 1976 the Prime Minister, Harold Wilson, took an interest.[119] The Council in the Isles of Scilly, where he had a holiday home, threatened to ban the import of all dogs, which led Wilson to worry whether he would be allowed to take his Labrador Paddy there that summer. The Prime Minister's Office made enquiries and was assured that because Paddy was a resident of the Scilly's any restriction would not apply to him.[120]

The summer of 1976 was the hottest on record in Britain, and saw in some ways a return to the fevered Dog Days of the nineteenth century, as, in the words of one correspondent, the country was gripped by 'rabies hysteria'.[121] The Ministry of Agriculture started things off with its usual publicity drive, which included new posters, more inspections, and encouragement to magistrates to use higher fines and prison sentences.[122] A one-minute warning film made by the Central Office of Information set out what was at stake.

> Can you imagine being frightened of every friendly animal you meet? Imagine rabies, in Britain. All dogs will be leashed and muzzled. Foxes will be destroyed. Wildlife at risk. No animal may be moved in or out of the infected area. All cats will be restrained. Just one animal – smuggled in – could lead to all this. So if you suspect anyone of smuggling, tell the police. If rabies breaks out, any animal found loose will be seized, taken away, and if it is not claimed, destroyed. Rabies is a killer. We must keep rabies out.[123]

In discussing its leaflet 'Rabies is a killer', a civil servant at the Information Division of the Department of Health wrote, 'we certainly don't want it over here' and went on 'it may sound like the Armada in history or an invasion of flying foxes in science fiction, but we are taking it seriously'.[124] Over the summer many smugglers were fined and in July two people were jailed for three months.[125] Criminal activity and graphic warnings made good copy for the popular press and produced deadlines such as, 'Rabies – It's 22 miles away' and the 'Hounds of Hell'.[126] However, a member of the RSPCA's governing council was reported as saying that the government was running a ' "terror campaign" to shock people, but it was inadequate, negative and likely to cause much suffering'.[127] He went on that though rabies was a horrific disease 'it could hardly be considered a plague' and that no one died of rabies in France for half a century.

The new Labour Prime Minister, James Callaghan, threw his weight behind the campaign and confirmed that the government was looking at making animal smuggling an arrestable offence.[128] The press kept the issue in the public mind, with good copy on cats being shot in Edinburgh, the slaughter of Fred an illegally imported mouse, dogs in handbags, and a couple were jailed for driving through customs with a sedated Afghan under a rug on the back seat of their car.[129] There was a Ministerial broadcast in May and experts wrestled with the problem of how to worry and reassure the public at the same time.[130] This was most obvious in accounts of rabies in France, where there were reported to be over 2,000 cases in animals each year.[131] Robert Adley, the MP for Folkestone, who regarded his constituents as on the front line, drew comparisons with Dunkirk, warning that because the Nazis stopped their advance at the Channel, it did not follow that rabies would too.[132]

The *Daily Mirror* reported, under the headline 'GUNS READY IN RABIES BATTLE', that local authorities had 'military-style plans for tackling every situation' and had been urged to have 'police marksmen standing by', with the Army as back-up.[133] Yet, the experience of tourists was that the only time they saw any warning about rabies was on British television or when re-entering Britain at the end of their holiday.[134]

Public complacency led some experts to suggest again that the protective vaccination of cats and dogs would be an alternative to quarantines.[135] A study by the Office of Health Economics concluded that the threat from rabies tended to be exaggerated in Britain and that vaccination was a viable and affordable option to quarantines, though its final recommendation was to maintain the status quo.[136] However, the government and it officials argued that vaccines were not foolproof, that it would be difficult to achieve the 75 per cent coverage necessary for protection, that such a scheme would be costly, and that it would harm the quarantine policy by creating a false sense of security.[137] Indeed, though it had a case, it seemed that Ministry of Agriculture, Fisheries and Food (MAFF) wanted to avoid any debate. One correspondent to the *Times* wrote that so little had been said about vaccination in Britain that she wondered if it had been the subject of a D Notice.[138] The case for vaccination was that it would be better to take precautions before the inevitable arrival of fox rabies, and that it would be better to plan to 'prevent' an outbreak, rather than 'cure' one with muzzling and slaughtering after it arrived.[139] Supporters also argued that vaccines were now effective and that 75 per cent coverage would not be hard to achieve. John Henwood, a veterinarian from Leicestershire wrote, 'No government has received more warning; unfortunately valuable time is being wasted and the country is not nearly as prepared as it could be'.[140] He was supported by Geoffrey Edsall, a former Harvard professor of public health, who maintained that British policy was contrary to WHO recommendations.[141] However, when the Chief Veterinary Officer entered the debate, he found support for government policy in another part of the WHO recommendations and argued that it was important to have a single policy and not to confuse the public. The central plank of the policy now seemed to be stigmatising and punishing 'the irresponsible act of animal smuggling'.[142] (Figure 6.4.)

In the late 1970s, the public face of anti-rabies policy remained unchanged, with the annual spring and summer government campaigns, port inspections, and the use of the courts to back up the warnings.[143] However, behind the scenes the Ministry and local councils drew up contingency plans for an outbreak, worrying about 'the most dangerous event for Britain ... an infected pet cat to be smuggled into the country and then escape into woodland'.[144] Such a scenario was written up in a fictional chapter, in an otherwise educational volume, *Rabies: The Facts You Need to Know* by G. N. Henderson and Kay White.[145] Their focus was a 'foolish' British family on holiday in France who decided to smuggle a stray cat back in their car. The narrator noted that, 'They thought that the import license and the quarantine regulations were just formalities that the government invents to make

Figure 6.4 'Don't Smuggle Death – Keep rabies out of Britain', RP4, Central Office of Information, 1977. This poster was designed for display within the United Kingdom and not only at ports.

life difficult'.[146] Once the family returned to England, its family dog, Toby, an award winning Cavalier King Charles Spaniel, licked up infected saliva from the smuggled cat, which then escaped into the countryside. Once the disease was recognised, the government implemented Rabies Orders and hundreds of pedigree dogs that had attended Cruft's, the country's premier dog show, were put under 'house arrest' in case they had been infected by Toby. Owners of suspect dogs were shown to have suffered terribly wondering about the fate of their pet and the guilt of the smugglers was deepened by a cull of wild animals in the 15 mile vicinity of their home.

The Henderson and White book endorsed MAFF policies and was designed to get the message through to the British public.[147] However, MAFF remained worried about overseas visitors, from diplomats to day trippers, who were ignorant, complacent, or both.[148] In 1978, one official was reported as saying, 'One tends to get the reaction that now we are in the Common Market, we should get everything they have got, including rabies'.[149] That the country was rabies-free and tough on rabies controls became an important feature of British identity, especially for the responsible classes who followed the quarantine rules. Of course, at the same time it raised questions about other features of Britishness, some of which Lord Ferrer captured when he asked people to report anyone seen flouting the regulations, 'This may seem a rather un-British thing to do. But it would take only one person smuggling one infected animal to introduce the horrifying disease into the country. Because we have not got it here, people in Britain say, "Oh, this is something that will not happen"'.[150] Such sentiments illustrate a key feature of national identity, that the English must never to drop their guard as foreign invasion is always a threat to their green, pleasant, and rabies-free land.

The menace of foreigners was clearly expressed in popular rabies fiction, a new genre which emerged in 1977 and 1978, with three major titles: *Saliva* (1977), *Return of the Mad Dogs* (1978) and *Rage* (1978). W. Harris's *Saliva* was saturated with allusions to the dangers posed by rabies and the Continent. The story begins showing the origins of a new rabies plague in badgers who, at the end of the Second World War had to feed on dead soldiers and in time developed a taste for flesh.[151] As the bodies run out, they start to attack the pets of holiday makers. Having told how rabies was transferred to domestic animals, the narrative moved to the political and diplomatic circles of Europe, where the disease spread amongst politicians who indulged their sexual appetites to stave off the boredom of life in Brussels. Rabies was remade as a sexually transmitted disease with moral meanings that distinguished between deserving and undeserving sufferers; hence, the book's subtitle – 'their lust let a killer into the country'. The Prime Minister's wife was the first British victim, having acquired it from her lover, Jean-Pierre Bouille, the Senior Secretary of the French Minister of Commerce; she first showed symptoms when collapsing on the stage at the Albert Hall.[152]

In Jack Ramsay's novel *Rage*, published the following year, sharp moral recriminations also take centre stage. A British politician, Lambert Diggery, imported rabies into Britain after allowing his daughter to bring in a dog that she had befriended whilst on holiday in France. Diggery was presented as a disreputable character, who enjoyed seedy relations with a prostitute in Brussels and even resorted to drug trafficking for kinky sexual favours. His moral bankruptcy was even more apparent when his child died from rabies. Unable to cope with the guilt, Diggery's wife confessed to the smuggling, while her adulterous husband resisted being implicated, seemingly not at all numbed by the death of his daughter. *The Day of the Mad Dogs* by David Ann followed the fate of two British tourists, John and Paula Denning, who smuggled a stray French dog into the country. Paula committed suicide, with a shotgun to her head, when she realised that she was developing rabies, while her husband, John, was kidnapped and tortured by a man seeking revenge after the death of his child from the disease. The novel had a dark apocalyptic climax, when the virus mutated into highly contagious air-born infection. The first victim of this new rabies, Lillian Shaw, a vivisectionist scientist, was gripped by a violent erotic mania and went on a hysterical rampage, seducing and sadistically killing men around her.[153] A mounting doomsday scenario was suggested with the very last line of the novel when Shaw declared this to be the first among many likely deaths. There is no evidence on how successful these novels were in terms of sales (they were very unlikely to have enjoyed any critical acclaim), there significance for us is what they reveal about popular images of rabies, their Europhobia and the ways rabies was transmogrified into a pandemic.

Through the late 1970s and early 1980s the nature and profile of rabies in Britain was defined by the annual government campaigns, sporadic press coverage of deaths from people bitten whilst abroad, and the sense that this was a foreign problem in every sense of the word.[154] It was ironic, therefore, that in the listings of human deaths from rabies in Europe, Britain was usually top because it had largest numbers of people travelling to and from South Asia, the region with by far the highest incidence of dog rabies in the world.[155] In 1978 and 1979 the situation in Europe started to change when the WHO reported that the spread of

rabies westwards had been halted.[156] This evidence was linked to observations in Continental Europe that the British government and its
people took rabies too seriously, and that fears were being needlessly
exaggerated for other reasons, such as justifying strict border controls
and anti-Common Market sentiments.[157] On the British side, it was suspected that Continental European governments played down the threat
to avoid alarming their citizens and to protect tourism.[158]

In 1982 reports showed that rabies in northern Europe was once again
on the increase, which led the British government's publicity machine
to move into overdrive on the issue, producing leaflets in 13 languages
as concern had switched from sentimental or ignorant British tourists
to blasé foreigners.[159] MAFF, local authorities and veterinary organisations continued to refine and rehearse their contingency plans for
containment should just one rabid cat, dog, or fox slip through quarantine.[160] The government line was that other EC governments took
rabies too lightly, began to be echoed in Brussels and amongst European
veterinarians, who started to discuss the possibility of eradicating the
disease in Europe. They were encouraged by the opportunities opened
up by new vaccines, produced by cell culture and other techniques,
which could be given orally to wildlife in bait.[161] While MAFF welcomed
the policy in principle, they were extremely troubled that this was
immediately linked to the scenario where, a rabies-free EC would have
common, community-wide regulations that would effectively allow the
free movement of all animals, in which case, Britain (and Ireland) would
no longer need their rabies quarantines.[162]

In 1983, the idea of rabies arriving in Britain was the basis for the
three-part primetime drama series, *The Mad Death*, produced by BBC
Scotland. Scripted by writer Sean Hignett, it was based on Nigel Slater's
novel of the same name. The drama followed the familiar idea of rabies
introduced into the country through a smuggled pet which, unbeknown
to its owner, had been bitten by a rabid fox. The action takes off when
the disease spreads to humans. The government mobilised veterinary
services and the army. Control measures were led by an experienced
veterinary officer, Michael Hilliard, working closely with a doctor – who
provided the love interest – to manage the public and avoid reactions
from complacency to panic. However, Hilliard's methods are seen by the
public as draconian. Pets were rounded up and vaccinated without their
owner's consent, while packs of wild dogs were chased across the
Scottish countryside and shot from a helicopter. Suspense was added
through the character of Miss Stonecroft, an eccentric upper-class spin-
ster, who lived in a large dark house that was home to a menagerie of

cats and dogs. In order to avoid Hilliard's controls, she deliberately released 60 dogs, which the army hunted down and killed, leading to a hostile reaction from the public. However, in the end the wisdom of Hilliard's tough methods were apparent when the outbreak was halted and the country returned to its rabies-free status, though the viewer was left with the warning that another similar, or perhaps worse, outbreak was inevitable.

The Mad Death was in part responding to fears that closer European integration was weakening Britain's border controls and national interest. The question of the harmonisation of regulations across the EC was the central issue at the political summit in December 1985, at which Britain and France 'came to blows over the notion of a Europe without frontiers'.[163] In the negotiations Britain's special position with rabies was used time and again to illustrate the folly of moves towards the free movement of everything. In a sense, the menace of rabies allowed Euroscepticism to be naturalised – a disease-free island next to an infected continent had to have special consideration.

In many EC debates, the need to protect the country from rabies was linked to other diseases (brucellosis, foot-and-mouth disease, and Colorado beetle) and to fighting terrorism, drugs, and illegal immigration, but often it was the sole, and assumed to be unanswerable illustration of Britain's claim for exceptions to rules that suited the rest of Europe.[164] Writing in the *Times* before the summit, Sarah Hogg warned that the proposals for integration may have gone too far, too fast, writing that, 'It is not hard to imagine the political mileage anti-Europeans could enjoy with the suggestion that the EC's latest wheeze would open Britain to a rabies epidemic.'[165] An editorial in the same issue argued that it was sensible to move to harmonisation in many areas, but not rabies and other similar issues.[166] And when Sir Geoffrey Howe, the Foreign Secretary, dismissed claims that Britain was seeking economic advantages from its opposition to reforms, he stated 'that Britain's natural concerns over rabies and drugs were not an excuse for protectionism'.[167] Norman Tebbit, a leading Eurosceptic Conservative politician and close ally of Margaret Thatcher, observed in 1992 that 'The blessing of insularity has long protected us against rabid dogs and foreign dictators alike'.[168]

From the Channel Tunnel to Pet Passports

The implications of a tunnel under the Channel for rabies had been discussed by British veterinarians as early as 1973, when, linking it to entry to the Common Market, they warned of the country being overrun with rabid dogs and foreign cattle breeds.[169] The fall-out from the 1985 EC summit coincided with the start of construction of the tunnel and MPs were repeatedly assured by Ministers that the contractors would observe strict precautions against rabies, such as 'sealed trains, physical barriers and grids at entry points, regular inspections, rigorous cleaning programmes, and the continuous deployment of baited traps in both tunnel and terminal areas'.[170] Over the period of construction, MPs wanted further assurances that rats and bats were being kept out, which led to the provision of strong, buried fencing and electrified grids.[171] The BBC political comedy series *Yes, Prime Minister* satirised French–Anglo relations at this time in an episode broadcast in December 1987. The comedy was provided by the French Prime Minister's attempts to undermine the British rabies policy by presenting a puppy to the Queen, even threatening to smuggle it into the country. British Ministers suspected the French government was trying to create a diplomatic incident to gain advantage in the Channel Tunnel negotiations; however, they won the day when British intelligence discovered that the French police had also brought with them a bomb to test British security. In October 1990 when the two sides drilling the Tunnel met, the *Daily Mirror* reported the event under the banner 'Channel link-up spells the end of our island race' and inevitably mentioned the special measures that would be taken to combat the threat of rabies.[172]

By the time the tunnel opened in 1994, the government was confident that it was no more dangerous than ferries, aircraft, or the many other ways that animals were brought into the country. However, Julian Barnes, a well known Francophile, wrote that on the day of its inauguration in May 1994, 'It was as if, lining up behind Mitterrand and the Queen as they cut the tricolour ribbons at Calais, were packs of swivel-eyed dogs, fizzing foxes, and slavering squirrels, all waiting to jump on the first boxcar to Folkestone and sink their teeth into Kentish flesh.'[173]

The American legal scholar Eve Darian-Smith has linked such hostile attitudes to Britain's post-colonial situation and fears of 'impending rapid change' in the country's 'spatial, legal and political relations with Europe'.[174] We would, of course, see these attitudes as longstanding and not necessarily linked to high-speed rail transportation or the end of Empire. Darian-Smith emphasised how Britain's rabies-free status 'reinforces a sense of the nation's unique superiority ... of a rational and law-abiding citizenry, as well as its ability to control its ports and borders.'[175] However, we see little evidence of the link she draws to the additional threat of importation of the disease by immigrants from former colonies in Asia and Africa, where she associates rabies with racial attitudes towards HIV-AIDS and Ebola virus. Where imported cases of rabies in humans occurred in the 1990s, they were seen by doctors and the press as unfortunate incidents, suffered by tourists and migrants. The only threat to health in Britain was to the staff caring for potentially violent patients. Rabies remained constructed, before and after the opening of the Channel Tunnel, as a problem of sentimental or ignorant tourists, European foxes, complacent European visitors, and enthusiasts for European integration, especially the Brussels bureaucracy.

By the mid-1990s the position with fox rabies in Europe had been transformed. Following many years of work, overseen by Commission veterinarians and with a 50 per cent subsidy, there was then talk of rabies being eradicated from the European Union (EU), the new name for the enlarged and reformed EC.[176] Julian Barnes might have been right about the British imagination, but on the ground reported cases of rabies in France had fallen from 2,500 in the mid-1980s to 200 in 1994; with most cases concentrated on the German border, a long way from Calais. By then it was much harder for British politicians to boast about the country's record on human–animal health because of mad cow disease (BSE – bovine spongiform encephalopathy) and the fact that Britain had been exporting diseased animals and meat around Europe for years. Nonetheless, the EU officials maintained that countries were on track to eradicate rabies by 1998 due to the application of vaccines used in successful Swiss campaigns in the late 1970s.[177]

As early as 1992 and buoyed by initial success, the Commission's Standing Veterinary Committee passed a resolution looking towards a community-wide, post-eradication policy of vaccination, blood tests, and the free movement of pets.[178] This proposal was swiftly condemned by the BVA, the RSPCA, the Quarantine Kennel Owners Association and MAFF, and was rejected in a motion passed by the House of Commons.[179] Government ministers argued that public opinion was behind quarantines and that they could only abandon the present system for one that gave even greater security. In fact, the BBC had shown a documentary in the First Sight series entitled 'Mad dogs and Englishmen' on 14 May 1992 in which the reporter argued that the quarantine laws gave reassurance to the public, but were not necessarily the best policy.[180] Ministers pointed to national peculiarities, which now included the large population of urban foxes. These lived at higher densities, were more sociable, and more liable to nuzzle and lick each other than the aloof rural foxes of the Continent. Models produced by epidemiologists projected that if rabies reached Britain, 92 per cent of foxes would need to be culled, or

95 per cent vaccinated, both were said to be impossible targets.[181] Officials pointed to a range of problems with any passport scheme: accuracy of the blood tests, out-of-date vaccinations, false passports, and the ease of 'impersonation'![182] An additional problem was that the government that had abolished the dog licence back in 1985 and would now have to make a U-turn and reintroduce a registration scheme.

Yet, by 1994 opposition to quarantines had begun to coalesce. The first, and in the end least influential, group to question the status quo was the House of Commons Agriculture Select Committee, which surprisingly included rabies in its enquiry into the health controls on the importation of live animals.[183] Two Labour members of the Committee, its Chair Jerry Wiggins and Dale Campbell Savours, were enthusiasts for reform and against most of the professional advice it heard, the Committee's report carried the unanimous conclusion that, 'scientific advances since the Waterhouse Report now made it not only feasible, but desirable, for the UK to permit anti-rabies controls based upon vaccination and blood testing as an alternative to quarantine'.[184] Members of the Committee, Labour and Conservative, seem to have adopted the ideas of Professor R. E. W. Halliwell, University of Edinburgh, and were also moved by the anguish of separation reported by many pet owners, and the likely scale of illegal imports that this was producing.[185] The Committee members were also hostile to MAFF, whose official lacked authority in the wake of salmonella and BSE, and welcomed evidence, such as that given by the animal behaviourist Roger Mugford, who observed that through government propaganda Britain had been 'infected with a phobia of rabies which is almost as interesting psychologically as rabies itself'.[186] In one question to the Minister, Gillian Shepherd, Gerry Wiggins also attacked the anti-rabies campaigns and commented sarcastically on the gap between British and European perceptions: 'The average Frenchman or the average German does not actually walk around with steel gumboots on being frightened of being bitten by a rabid fox and does not understand why we are so frightfully worked up about rabies'.[187]

The Committee's recommendations were ignored by the government and their cause was not helped when the British Medical Association and BVA also came out against any change of policy in July 1995.[188] However, the authority of the BVA was challenged in 1996 by a group that called itself 'Vets in Support of a Change', which included Richard Halliwell and high-profile veterinarians, such as the Conservative peer Lord Soulsby, and the broadcaster Bruce Fogle.[189] Their manifesto, which called for vaccination, blood testing and certification, argued that the 'present laws are anachronistic, and are indefensible on scientific grounds'. They also called upon their veterinary colleagues to show they were open-minded and to embrace change.

The most important motor of change turned out to be the lobby group 'Passports for Pets'. This was founded in 1994 by Lady Mary Fretwell, along with her husband Lord Fretwell who was a former British Ambassador to France. In fact, the term 'Pet Passports' had been coined by Screaming Lord Sutch of the Monster Raving Loony Party in the 1980s, though his scheme was about the ridiculous rather than rabies. The Fretwell's campaign gathered momentum in 1996 when Chris Patten, a leading Conservative and then Governor of Hong Kong, started to worry about putting his two Norfolk terriers, Whisky and Soda, in quarantine on his return to Britain in 1997 when the colony was taken back by China. Influential columnists, such as Simon Jenkins and Brian Sewell, published articles in support of a change, arguing that the current policy only served the kennel lobby.[190] An article in *The Independent* asked pointed, 'Why aren't Germans all dead from rabies?'[191] Press reports in November 1996 stated that John Major was minded to change government policy, but it was omitted from the Conservative election manifesto the following year.[192] In fact, Patten was so disgusted with quarantines that he sent his dogs to his holiday home in France rather than have them placed in quarantine. Mrs Patten stated in 2000 that,

> We never contemplated leaving them behind in Hong Kong. We thought about quarantine but we didn't want to go through with it. So we put them on a flight to Paris. We then went to France for nine months while Chris wrote his book. We have been waiting since then for the law to change so that they could come back with us. The dogs have both been very happy. It has been me who has been missing them.[193]

New Labour included a review of quarantines in its manifesto for the 1997 election and committed to the policy within six weeks of coming to power. By this time, Passports for Pets and Vets in Support of a Change had been joined by the RSPCA, and together they hired the health lobby firm Lowe Fusion to orchestra a media and public campaign under the umbrella of the Quarantine Reform Campaign (QRC) (Figure 6.5).[194]

The choice of Lowe Fusion was no accident. It was part of the Lowe Group, a company started in 1981 by Frank Lowe in London and which

Figure 6.5 Quarantine Reform Campaign. Reproduced by permission of Lady Mary Fretwell

by the end of the 1990s was the 14th largest advertising company in the world. The *Daily Telegraph* reported in September 1998 that 'Frank Lowe, founder and chairman of the advertising firm, the Lowe Group, still lives in Switzerland because he cannot contemplate quarantine in Britain for his Pekingese Lucy and William.'[195] Even allowing for the element of self-promotion, Lowe Fusion's description of its work for QRC is interesting.

> This initiative was carried out by the agency on behalf of a number of interested parties ... which all joined forces under the Quarantine Reform Campaign (QRC) banner. To support QRC's lobbying activities, Lowe Fusion set to work to generate public pressure and to help key political decision-makers understand and accept the need for a review of the existing quarantine regulations. The agency achieved this by generating a number of pro-active and re-active media stories. The support of various pet owners was enlisted, especially those who could give a personal insight into the conditions of the quarantine kennels and the restrictions some owners endured by not being able to travel with their animals. With such a notoriously emotive issue as rabies, Lowe Fusion worked tirelessly to produce a full media pack detailing the activities and objectives of the QRC with case studies and contacts for comment. On behalf of the QRC, the agency also published a 12-page report, The Case for Change, designed for a non-scientific audience. Sir Richard Branson was just one of many celebrities and key media personnel who agreed to support the campaign leading up to its high-profile launch at the Palace of Westminster.[196]

QRC activities included lobbying the Labour Party Conference in 1997, a petition handed to the Prime Minister in July 1998, and a protest at Parliament in October 1998.

However, change was already afoot. By this time fox rabies had been almost been eliminated from EU countries; already in 1997 an article in the *New Scientist* was actually speculating on what conditions might lead it to return! The journal's Brussels correspondent, Debora MacKenzie, reported, 'No cases have been detected for a year in France, and for almost a year in Switzerland and Belgium. The Netherlands, Luxembourg and the former East Germany have had no cases for two years and can officially declare themselves rabies-free'.[197] The remaining infected areas were in Germany, where the problems were the political ones of the will and money to complete the task, there were seemingly no issues with the science or technology of eradication.[198] It was against this background, which had been well articulated in Parliament and press, that on 2 October 1997, the Minister of Agriculture Nick Brown announced the details of the promised enquiry, chaired by Professor Ian Kennedy, an academic lawyer and medical ethicist, to assess alternatives to quarantine. The Group's deliberations and report were highly technical and focused on risk assessments, but ended up recommending radical changes, albeit implemented in a guarded fashion.[199] While the Group worked, lobbying for change did not relent, the QRC continued to enrol celebrities, such as Richard Branson, Jilly Cooper, Liz Hurley, David Hockney, Chris Patten, and Sting; to keep up the moment in Parliament with questions to Ministers; and to keep the issue in the media.[200] The documentary *Inside Quarantine* in Channel 4's 'Undercover Britain' series, shown in May 1998, which revealed lax and unsanitary conditions in one kennel helped undermine public faith in the reliability and effectiveness of the quarantine system.[201]

The Kennedy Group's Report *Quarantine and Rabies: A Reappraisal* was published in September 1998.[202] It recommended scrapping the six-month quarantine on animals entering from the EU, Australia, New Zealand, and Hawaii, if pets had a certificate of vaccination and blood tests confirmed that it had been effective. Such animals would be identified by a microchip and, because of the anticipated increase in pet traffic; it was recommended that dogs also be required to be treated for two parasites – the fox tapeworm and the kennel tick.[203] Nick Brown responded immediately by announcing that the government accepted the recommendations and would move to introduce 'Pet Passports'.[204] He received cross-party support and the government began to look at the similar scheme that had operated in Sweden since 1994.[205] The details of the new scheme were announced in March 1999, with the commitment for them to be operational within 12 months.

The Pet Travel Scheme (PETS) was introduced on 28 February 2000 and was front page news.[206] In fact, the first passport had been issued by the Prime Minister in August 1999, to a boy who was terminally ill so that he could have his pet with him during his final days.[207] This flexible response to a human tragedy was a poignant introduction and in marked contrast to the rigid, uncompromising attitudes that had characterised quarantine regulations for a century. Given the extent of united veterinary, medical, and political opposition barely five years before, it was remarkable that the new scheme was introduced with hardly a word of protest. Moreover, the public seems to have had complete trust in vaccinations and microchips, as there was hardly any mention in the publicity or news reporting of the eradication of rabies from

Continental Europe.[208] The change was a great success for the Passports for Pets campaign, achieving an objective that when the organisation was founded would have seemed 'Mission Impossible'.

Conclusion

Screaming Lord Sutch was famous for his absurd policies, so what turned a Raving Loony policy into a New Labour policy? PETS was very New Labour. It was seemingly a response to pressures from the middle and upper classes with properties abroad, to corporate interests, to celebrity endorsements from pop and movie stars, and to the actions of political insiders like the Fretwells and Pattons. There was apparently no demand from Labour's heartlands, the trade unions, nor the wider pet owning public. Pet Passports also signalled a new attitude to Europe, where the distrust, suspicion, and exceptionalism of the Tory years was abandoned in favour of liberty, fraternity, and equality for dogs.

However, there were also technocratic and social factors working for the scheme. The retreat of rabies in Europe, made possible by the application of vaccination technologies, had been supported by the EU and meant that any threat from European fox rabies was small. There were also new forms of state power in play, as surveillance, monitoring, and supervision signalled a shift away from the authoritative displays, evident in quarantines, fines, and prison. People were travelling more, aided by rising wages, cheaper fares and were buying property abroad. People were less willing than ever before to leave their pets at home or tolerate a long quarantine, seemingly, they were more sentimental than ever over their pets. The growth of small animal veterinary practices, along with pet insurance and health care systems, meant that owners invested more and more financially and emotionally in their pets. However, PETS may have been deliberately complex and expensive. The scheme allowed entry to micro-chipped dogs and cats with effectively proven vaccinations, shown by a certificated blood antibody test. Quarantines remain for those without the resources or foresight to plan the necessary repeat visits to veterinarians to qualify, and for those 'foreigners' who remained unaware of Britain's superior status as a rabies-free nation.

Conclusion

Noel Coward's mad dogs that went into the mid-day sun were foolish rather than ferocious and deadly. When he wrote the song in the early 1930s, rabies was no longer the threat it had once been, rabid dogs had been banished and the human disease was routinely treatable. Britain had yet to experience the growth of mass overseas tourism that would lead the government and veterinary experts to re-establish public fears over the importation of the disease in the 1960s and 1970s. It is paradoxical that Coward's phrase remains so resonant when he draws upon such a singular image of mad dogs. Before and since the 1930s, the English were characterised as having greater fear of rabies than almost any other nationality. One reason for this was that rabies in Britain had been like Cerebus – the many headed dog in Greek mythology – with multiple identities and meanings. In the Introduction we said that we did not want to set out what rabies *is*, but wanted readers to discover what rabies *was* along with our historical actors. What we have revealed, across two centuries, is that rabies *was* constructed as many things and that its many forms were made and remade by its many 'experts'.

Mad dogs and rabies

Some constructions of rabies were enduring and others quite specific to time, place, and social location. To veterinarians in the 1820s and 1830s rabies was *canine madness*, a disease found in individual dogs and passed on by the bite of an infected animal. However, the term also referred to a condition of the dog population, which was the result of neglect and ill-treatment by the poor and symbolised the wider social and political crisis. To doctors rabies was *hydrophobia*, its human form caught from a rabid dog bite. But were rabies and hydrophobia the same? Some doctors argued that they were different diseases: humans became violently averse to water whereas dogs showed great thirst. Some medical men were convinced that there was a second type of human disease – *spurious hydrophobia* that could be brought on by anxiety. Also well recognised by local elites and social reformers was what can be termed *mad dog rabies*, a street phenomenon when the call of 'Mad Dog' could produce spontaneous disorder and violence, as the veneer of civility broke down and innocent dogs were brutally killed by unruly mobs.

In the 1840s anticontagionist veterinarians and doctors created *zymotic rabies*, a disease analogous to cholera and other filth diseases, especially in the sense that it could be spontaneously generated as well as spread by contagion. This form was favoured by animal welfare reformers who identified rabies with strays, curs, and the working-class culture of dog pits, dog carts, and neglect of man's best friend. The construction had greatest currency in the 1840s and 1850s when the reported prevalence of rabies was quite low and when animal welfare reformers enjoyed some success in welfare and police legislation. They were convinced that greater humanity to dogs would suppress the problem. Increased incidence in the 1860s saw veterinarians remake the condition as *epizootic rabies* – an imported animal plague – analogous to the cattle plague and other highly contagious animal diseases. At the same time, veterinarians made clearer distinctions between *furious rabies* and *dumb rabies*, linking each to the character, training, and breed of a dog. The fact that veterinarians elaborated accounts of a disease that was so variable in its development, symptoms, forms, and consequences did little to help their authority. George Fleming in particular recognised that he had to confront popular understandings in order to make the disease amenable to control.

In the 1880s Pasteur brought rabies into the laboratory and through experimental manipulation produced a new, stable, and replicable disease with a six-day incubation period. His antivivisectionist critics called his new disease *laboratory rabies*, an affliction that summoned up images of Frankenstein and which they claimed was more dangerous and deadly than natural rabies. Scientists also adopted this term and in turn created *street rabies* for the natural, variable form that everyone was familiar with. Pasteur disciplined the virus so that he could produce and reproduce the disease at will; he also made rabies a treatable disease. In addition, he created a standard test that showed rabies had the same cause in all species, which signalled the beginning of the end for the term hydrophobia. Indeed, by the early twentieth century doctors and scientists were only writing of the singular *rabies* that is familiar to us today: a disease of mammals that is spread by the inoculation of a virus that moves through nerve cells to the spine and brain, to produce encephalitis and rapid death. This *rabies* has remained stable in science and medicine for nearly a century, albeit refined by greater knowledge of the virus, its actions, and the body's immune reactions.

In the twentieth century new versions of rabies were created by governments and their agents. In Britain the disease became *foreign*, associated first with Continental Europe and then with tropical, third world countries. Its exotic character was reinforced by the identification of new forms discovered in vampire bats and European foxes; indeed, in the modern British consciousness the disease is mostly *fox rabies*, a form spread by a sly scavenger that poses a particular threat to pets and children.

We have identified many constructions rabies and could produce more by including metaphorical as well as literal uses. Our main point, however, is that there was not a single narrative of the rabies in Britain, but many forms shaped and sustained by its various experts who interacted with each other, and whose views were also shaped by the behaviour of dogs and the virus, along with social and political factors. Many of the constructions of rabies we have identified were complicated further by gender, breed, and class. Male dogs were assumed to be more susceptible than bitches, with sexual frustration a widely canvassed cause of spontaneous origins in the nineteenth century. Certain breeds were believed to be more susceptible than others; street curs – the dogs of the working class, that were occasionally called 'working class dogs' – were seen as the main sources and carriers of rabies through to the present day.

Englishmen, Englishwomen and rabies

Noel Coward's Englishmen in the tropics would have thought 'Mad Dogs' were anything but foolish. They would have found them just as terrifying as their nineteenth century domestic forebears, probably more so, as dogs in colonial towns and villages would have seemed wilder and more threatening. Amongst Englishmen and Englishwomen at home there were five main groups of 'experts' on rabies: veterinarians, doctors, the public, animal welfare reformers, and state officials, both functionaries and politicians. We might expect veterinarians to have been the most authoritative group given that rabies was primarily an animal disease. However, sustained professional practice with dogs did not develop until the second half of the twentieth century and before then 'dog doctor' was a term of abuse. In fact, expertise on rabies in the nineteenth century has been attributed by veterinary historians to three men: Delabere Blaine, William Youatt, and George Fleming. However, their consistent view that it was a contagious disease was contested by other veterinarians, doctors and lay experts, and disagreements amongst veterinarians further undermined the authority of a relatively weak profession. Things changed after 1897, when an alliance of politicians, state officials, and veterinarians promoted a consistent contagionist view which dominated policy and legitimated strict surveillance and controls for a century.

Over the period covered in this book, medicine was the higher-ranking profession and while doctors regularly wrote on animal diseases, they rigorously excluded veterinarians' incursions into human disease. However, there was also no consensus in medicine about hydrophobia until the late nineteenth century. The sporadic nature of the disease and the fact that victims died so quickly meant that there was little opportunity for doctors to gain sustained experience in any aspect of its nature. The disease itself was unhelpful as there were no obvious morbid changes observable at post mortem and attempts to isolate the poison failed until Pasteur's investigations in the 1880s. Then, rabies became one of the first diseases to be reshaped by the new germ theories of disease, even though no bacterium was revealed by microscopy or plate culture. Through the twentieth century medical knowledge was relatively stable, maintained by a relatively small group of virologists, few of whom worked in Britain.

Mad dog incidents – the public spectacle of chases, killings, and controls – were more common that veterinary or medical cases of rabies. The drama and the emotions they raised kept them in the popular memory and public imagination. In this sense, ordinary Englishmen and women had greater experience of rabies than professional experts; they were able to draw upon the collective expertise of community traditions and folk law about killing dogs and treating victims. Doctors dismissed popular nostrums, like treating with the hair of the dog and the remedies peddled by local chemists, but these practices endured, not least because until Pasteur's work doctors were seen to offer nothing beyond preventive surgical treatment. The continuing hold that rabies had on the emotions and the popular imagination was also evident in fiction and in metaphors drawn from knowledge of the symptoms in humans and in dogs.

Throughout the nineteenth century animal welfare reformers argued that the origins of rabies lay in cruelty and neglect. Their analysis reflected larger preoccupations with the reform of working class culture and the promotion of humanitarianism. Fear of rabies, especially its spontaneous eruption, was mobilised in campaigns to ban dog fighting and dog carts, and to rid the streets of dangerous dogs and strays. The latter proved to be an intractable problem and seemed to worsen in the late Victorian period when tolerance of rowdy street cultures diminished, and when the working class cur became a sentimentalised domestic pet as well as a creature of the street. By this time animal welfare campaigners had changed their tune, from arguing that rabies was potentially everywhere due to cruelty, to the view that the threat had been overstated. They now pointed to the low number of human deaths, and emphasised just how rare proven rabies cases actually were.

Animal welfare reformers linked popular violence to that perpetrated in laboratories by scientists, though with the qualification that the cruelty of the professional middle class vivisectionists was worse because it was calculated and knowing, rather than the spontaneous and ignorant cruelty of the working class. Antivivisectionists claimed that rabies scares were the inventions of scientists like Louis Pasteur and Victor Horsley, who used them to justify their inhuman experiments and dangerous cures. That leading scientists also advocated universal muzzling confirmed reformers' views of their inhumanity. The defenders of dogs, who were increasingly women, argued that muzzles were ineffective in the fight against rabies and would perpetuate the disease by maddening dogs, making well-cared for dogs vulnerable, and giving a false sense of security. In the twentieth century, when rabies had been stamped out and was being kept out, the voice of animal welfare reformers was still heard, not least over the inhumanity of quarantines. With the mounting criticism of quarantines towards the end of the twentieth century, the groups that came together in the QRC argued that the unhappiness and stress in dogs and their owners, induced by separation, produced greater suffering than rabies was ever likely to cause, especially given the availability of effective vaccines.

We have shown that in the early nineteenth century the state's role in rabies control was minimal and permissive. Attempts to introduce specific legislation in 1830–1831 failed; there was little success in subsequent decades and responsibility for control measures remained with local authorities which used a variety of legal instruments. This situation continued for the best part of the Victorian period. It was not until 1886, when rabies was added to the Contagious Diseases (Animals) Act that central government took a leading role, though its Veterinary Department initially pursued suppression rather than stamping out. The aim of eradicating rabies was not adopted until 1897 when Walter Long made it a symbol of the strong state advocated by Tory Unionists. His success with the policy shored up the notion of rights of the state over the individual in certain spheres; the fact that this was done in the face of enormous political protest against a climate of social conflict reinforced a political vision of unity and cohesion. In the twentieth century, rabies was kept at out by the Veterinary Department of the Ministry of Agriculture, which created a system of surveillance and border checks that criminalised dog smugglers. Indeed, in government propaganda, dog smugglers were portrayed as potential mass murderer of wildlife, pets, and perhaps even their fellow citizens. From the 1970s, some politicians made rabies quarantines symbolic of the character of an island people and the superiority of its state apparatus. This evocative image of a strong nation fed into wider distrust of Europe, where rabies seemed easily to sweep across borders and where little seemed to be done to halt its deadly advance. Keeping rabies out was a battle about political vision, the future of the nation, and even its sovereignty.

At the turn of the twenty-first century New Labour introduced Pet Passports, a move that signalled a new kind of state and a new kind of relationship with Europe. Gone was the xenophobia of previous rabies policies and in its place was a policy that trusted European institutions and peoples. Pet Passports also ushered in a new vision of the state that no longer depended on rigid and draconian displays of power in the name of British sovereignty. Proclamations about Pet Passports embodied New Labour's emphasis on citizenship that construed a positive relationship between the state and the individual. However, the scheme harboured contradictions. Obtaining a Pet Passport was expensive and complex. It seems to have been designed for pet owners with overseas properties, for celebrities, and for the wealthy who were regularly in and out of the country. The scheme was not designed for the ordinary holidaymaker, and quarantines remained in place for the unprepared and uninformed. Moreover, surveillance, fines, and the threat of prison remained in place to deter 'smugglers', be they a family whose child befriended a stray pet, or the sentimental pet owner who at the last minute could not bear to be separated from their beloved pet, or the determined importer of banned dog breeds. In other words, strict controls remained for those Englishmen and Englishwomen that the state, animal welfare reformers, and professionals had always associated with mad dogs.

Notes

Introduction

1. Until the 1970s the primary concern in Britain was rabies in dogs and cats. Only in the last 20 years of the twentieth century did policies shift to the threat from the disease in wildlife, first foxes and latterly bats. See Chapter 6.
2. There are few histories of rabies, but see J. Théodordiès, *Histoire de la Rage: Cave Canem*, Paris: Masson, 1986; L. Wilkinson, 'History', in A. C. Jackson and W. H. Wunner, eds, *Rabies*, Amsterdam: Rodopi, 2002, 1–22.
3. M. Worboys, 'Vaccines: Conquering Untreatable Diseases', *British Medical Journal*, 2007, 334(suppl_1): 19.
4. *WHO Expert Consultation on Rabies: First Report*, WHO Technical Report Series 931, Geneva: WHO, 2004, 13; B. Dodet, 'Preventing the Incurable: Asian Rabies Experts Advocate Rabies Control', *Vaccine*, 2006, 24: 3045–49.
5. C. E. Rosenberg, 'Framing Disease: Illness, Society and History', in C. E. Rosenberg, and J. Golden, eds, *Framing Disease: Studies in Cultural History*, New Brunswick, NJ: Rutgers University Press, 1992, xiii–xxvi; M. Worboys, *Spreading Germs: Disease Theories and Medical Practice in Britain, 1865–1900*, Cambridge, MA: Cambridge University Press, 2000, 8–14; C. E. Rosenberg, 'What is disease? In Memory of Owsei Temkin', *Bulletin of the History of Medicine*, 2003, 77: 491–505.
6. Up to date information can be found on the websites of the World Health Organisation http://www.who.int/topics/rabies/en/ and the European Commission. Also see M. J. Warrell and D. A. Warrell, 'Rabies and Other Lyssavirus Diseases', *Lancet*, 2004, 363: 959–69.
7. A. R. Fooks, D. H. Roberts, M. Lynch, P. Hersteinsson and H. Runolfsson, 'Rabies in the United Kingdom, Ireland and Iceland' in A. A. King, A. R. Fooks, M. Aubert and A. I. Wandeler, eds, *Historical Perspective of Rabies in Europe and the Mediterranean Basin*, Paris: l'Organisation mondiale de la santé animale (OIE), 2004.
8. L. Wilkinson, *Animals and Disease: An Introduction to the History of Comparative Medicine*, Cambridge, MA: Cambridge University Press, 1992; A. Hardy, 'Animals, Disease and Man: Making Connections', *Perspectives in Biology and Medicine*, 2003, 46: 200–15.
9. J. H. Bell, E. Fee and T. M. Brown, 'Anthrax and the Wool Trade', *American Journal of Public Health*, 2002, 92: 754–57; L. Wilkinson, 'Glanders: Medicine and Veterinary Medicine in Pursuit of a Contagious Disease', *Medical History*, 1981, 25: 363–82; Psittacosis, which can cause severe pneumonia, is caused by the bacterium *Chlamydophila psittaci* and is caught by pet shop owners and bird fanciers.
10. Worboys, *Spreading Germs*, 63–9, 150–61, and 221–27.
11. A. Hardy, 'Pioneers in the Victorian Provinces: Veterinarians, Public Health and the Urban Animal Economy', *Urban History*, 2002, 29: 372–87.

12. H. Koprowski and S. A. Plotkin, eds, *World's Debt to Pasteur: Proceedings of a Centennial Symposium Commemorating the First Rabies Vaccination*, New York, NY: A. R. Liss, 1985, 141–218.
13. Epizootic and enzootic are the veterinary equivalents to epidemic and endemic diseases with human disease. An epizootic is a disease that is not normally present or very prevalent in an area, but which is imported or increases its prevalence. An enzootic disease is one that is normally present, being linked to the social, economic, biological or physical geography of the area.
14. M. Brock, *The Great Reform Act*, London: Hutchinson, 1973.
15. J. K. Walton, 'Mad Dogs and Englishmen: The Conflict Over Rabies in Late Victorian England', *Social History*, 1979, 13: 219–39.
16. Our inspiration and exemplar for this approach is, of course, Roy Porter, especially: R. Porter and G. S. Rousseau, *Gout: The Patrician Malady*, New Haven, CT: Yale University Press, 1998.
17. K. Thomas, *Man and the Natural World: A History of Modern Sensibility*, London: Allen Lane, 1983.
18. H. Ritvo, *The Animal Estate: The English and Other Creatures in the Victorian Age*, Cambridge, MA: Harvard University Press, 1987.
19. Ritvo, *Animal Estate*, 170.
20. Darwin Correspondence Online, Letters: 1211, 4 February 1849; 1904, 17–18 June 1856; 3394, 18 January 1862; 3792, 3 November 1862; George Eliot, *Mill on the Floss*, Penguin Classics, 1992. First published in 1860. Ch 12 and Anthony Trollope, *The Eustace Diamonds*, Ch. 57, London: Chapman and Hall, 1873.
21. *American Historical Review*, 1989, 94: 771; *Journal of Interdisciplinary History*, 19, 1988: 323.
22. J. Burt, 'The Illumination of the Animal Kingdom: The Role of Light and Electricity in Animal Representation', *Society and Animals*, 2001, 9: 203. Also see: Erica Fudge, *Animal*, London: Reaktion Books, 2006.

1 Rabies Raging: The 'Era of Canine Madness', 1830

1. National Archives, Kew. HO 44/20, 363–4. Letter from 'Polydius' to Sir Robert Peel, June 1830.
2. HO 44/20, 195. Samuel Bardsley (Manchester) to Sir Robert Peel, 24 May 1830.
3. *Morning Advertiser*, 23 June 1830, 3d.; *Manchester Courier*, 15 May 1830, 4c.
4. M. D. George, *Catalogue of Political and Personal Satires in the British Museum*, Vol. 10, London: British Museum, 1952, [15360], 649.
5. *Evening Mail*, 7 May 1830, 6b.
6. *Times*, 4 June 1830, 3d.
7. 44/20 236–7. Letter from Curtis to Robert Peel, 2 June 1830.
8. HO 44/20. Letter from Scotics to Robert Peel, 3 June 1830.
9. R. Powell, *Dr Powell's Case of Hydrophobia*, London: G. Woodfall, 1808; B. Moseley, *On Hydrophobia, its prevention and cure*, London: Bensley, 1808. On rabies in the late eighteenth and early nineteenth centuries see J. D. Blaisdell, *A Frightful – But Not Necessarily Fatal – Madness: Rabies in Eighteenth Century England and English North America*, Unpublished PhD Thesis, Iowa State University, 1995.
10. S. Butler, 'Bardsley, Samuel Argent (1764–1850)', *Oxford Dictionary of National Biography*, Oxford University Press, 2004 [http://www.oxforddnb.com/view/

article/1363, accessed 10 March 2007]. S. A. Bardsley, *Medical Reports of Cases and Experiments, with Observations, Chiefly Derived from Hospital Practice: to Which are Added an Enquiry into the Origin of Canine Madness; and Thoughts on a Plan for its Extirpation from the British Isles*, London: Printed by W. Stratford, for R. Bickerstaff, 1807; J. M. Stratton, *Samuel Bardsley and the History of the Prevention of Rabies*, Croydon: H. R. Grubb, 1960; S. A. Hall, 'The Bardsely Plan and the Early Nineteenth Century Controversy on Rabies', *Veterinary History*, 1977, 9: 15–21.

11. On the use of gunpowder see H. Sully, *Observations on, and Plain Directions for all Classes of People to Prevent the Fatal Effects of the Bites of Animals Labouring Under Hydrophobia*, R. Hall: Taunton, 1828, 40–41. On amputation see: J. Elliotson, 'Lectures on the Theory and Practice of Medicine: Hydrophobia', *London Medical Gazette*, 1833, 12: 505.

12. W. Lawrence, *Times*, 25 July 1830, 7a–c.

13. Elliotson reported many other medical measures, for example, the whirling chair in asylums, narcotics of various types, and mercury. Elliotson, 'Lectures', 504.

14. W. R. Hunter, 'William Hill and the Ormskirk medicine', *Medical History*, 1968, 12: 294–7.

15. J. Murray, *Remarks on the Disease called Hydrophobia; Prophylactic and Curative*, London: Longman, Rees, Orme, Brown and Green, 37–38.

16. Elliotson, 'Lectures', 504.

17. *Medical Gazette*, 1830, 6: 411.

18. J. Elliotson, 'Clinical Lecture: Hydrophobia', *Lancet*, 1830, 15: 284–92.

19. W. Simpson, 'Case of hydrophobia', *Lancet*, 1831, 17: 29.

20. Elliotson, 'Clinical', 288.

21. W. Lawrence, *Times*, 23 July 1830, 7a–c.

22. C. Brady, 'Case of Hydrophobia', *Lancet*, 1829, 12: 341.

23. Ibid.

24. A. T. Fayerman, 'Case of Hydrophobia', *Lancet*, 1825, 3: 25.

25. J. D. Blaisdell, 'Rabies and the Governor General of Canada', *Veterinary History*, 1992, 7: 19–26.

26. Murray, *Remarks on the Disease*, 344–6.

27. A. Scull, *The Most Solitary of Afflictions: Madness and Society in Britain, 1700–1900*, London: Yale University Press, 1993.

28. J. Connolly, 'Symptoms Following the Bite of a Dog Supposed to be Rabid', *Veterinarian*, 1832, 5: 95.

29. HO 44/18, 51–58, 55. Letter from Dr Henry Thompson to Sir Robert Peel, March 1828. Underlining in original.

30. The topic was aired frequently in the columns of *Notes and Queries*. See *Notes and Queries*, 1852, Series 1, 6: 206–7, and 298–99; Series 2, 1: 362 and 442; Series 4, 10, 1872: I, 382, and 430–40; Series 10, 1: 65, 176, 210, and 332.

31. J. Bardsley, 'Hydrophobia', in J. Forbes et al., eds., *The Cyclopaedia of Practical Medicine*, London: Sherwood, Gilbert and Piper, 1833, 483–518, 483. On popular attitudes to the medical profession around dissection and the passage of the Anatomy Acts see R. Richardson, *Death, Dissection and the Destitute*, London: Routledge, Kegan and Paul, 1988.

32. Elliotson, 'Lectures', 501.

33. *Committee on Bill to Prevent Spreading of Canine Madness Report*, Minutes of Evidence, Parl. Papers, 1830, (651), x, 703.

34. R. White, *Doubts of Hydrophobia as a Specific Disease to be Communicated by the Bite of a Dog, with Experiments on the Supposed Virus Generated in that Animal during the Complaint termed Madness*, London: Knight & Lacey, 1826, 39.
35. J. Pattison, *The British Veterinary Profession*, London: J. A. Allen, 1984. The discussion that follows is mainly based on Pattison's review.
36. The London Veterinary College, founded in 1791, took on the title of the Royal Veterinary College in 1826 due to the patronage of George IV. London Veterinary Medical Society started meeting in 1813 and was succeeded by Veterinary Medical Association in 1836. In Scotland, William Dick began lecturing on 'the diseases of horses, black cattle, sheep and other domesticated animals' in 1823, thereby founding the Dick School. The first national journal *The Veterinarian* was founded in 1828 by William Percivall, and The Royal College of Veterinary Medicine received in Charter in 1836, acquiring at least formally the same status as the Royal Colleges in medicine.
37. D. Blaine, *Canine Pathology*, London: T. Boosey, 1817.
38. Ibid., 96–132.
39. Ibid., 98, 124.
40. Ibid., 103.
41. W. Youatt, *On Canine Madness Being a Series of Papers Published in 'The Veterinarian' in 1828, 1829, and 1830*, London: Longman and Co., 1830.
42. H. W. Dewhurst, *Observations on the Probable Causes of Rabies, or Madness in the Dog, and Various Other Domestic Animals*, London: Alexandre, 1831.
43. W. Youatt, 'On Rabies Canina', *Veterinarian*, 1828, 1: 30–31. The lecture series was published in eight instalments. Ibid., 28–32, 62–4, 139–41, 153–9, 281–5, and 1829, 2: 86–90.
44. Ibid., 281.
45. Ibid., 157.
46. W. Youatt, 'On the Communication of Rabies by all Rabid Animals', *Veterinarian*, 1831, 4: 230.
47. In fact, there were many reports from French veterinary schools with often varying results that were used to produce contradictory conclusions. There was also some experimental work in Britain, notably by John Hunter and Hugo Meynell. Blaine and Youatt also experimented on the disease. L. Wilkinson, 'John Hunter and the Transmissibility of Rabies', *Veterinary History*. 1981–1983, 2: 78–84; J. D. Blaisdell, 'John Hunter (1728–1793) and Rabies' *Veterinary Heritage*, 1989, 12: 19–37.
48. Committee on the Bill, 687. This was also the position of Thomas Southwood Smith, 'Hydrophobia', *Westminster Review*, 1824, 2: 324–34.
49. 'Report of Meeting of the Veterinary Medical Society, 6 October 1830', *Veterinarian*, 1830, 3: 659.
50. T. J. Pettigrew, *Substance of a Clinical Lecture on a Case of Hydrophobia, Delivered at the Charing Cross Hospital ... to Which are Appended the Particulars of Another Case*, London: Longman, 1834. Pettigrew also wrote a manuscript for a book on hydrophobia that was never published, perhaps because of the declining incidence of the disease in that decade. Pettigrew Collection, Collection of notes, extracts, and so on used for his work on 'Hydrophobia', Wellcome Library, Western Archives and Manuscripts, MS3864 and 3865. The item is described in S. A. J. Moorat, *Catalogue of Western Manuscripts on Medicine and Science in the Wellcome Historical Medical Library*, London: Wellcome Institute for the History of Medicine, 1962–1973.

51. M. Harrison, 'From Medical Astrology to Medical Astronomy: Sol-lunar and Planetary Theories of Disease in British Medicine, c. 1700–1850', *British Journal for the History of Science*, 2000, 33: 25–48.
52. HO 44/14. Letter from Henry Brandon to Sir Robert Peel, 27 June 1825.
53. Youatt, 'On rabies canina', 281–85.
54. Dewhurst, *Observations on the Possible Causes of Rabies'*. This book was based on a lecture read at the London Veterinary Medical Society, 6 October 1830.
55. *Veterinarian*, 1830, 3: 283.
56. HO 44/18. Letter from Dr Henry Thompson to Sir Robert Peel, March 1828, 55.
57. *Veterinarian*, 1829, 2: 86.
58. Ibid.
59. *Times*, 18 April 1825, 6e.
60. *Times*, 13 July 1825, 3b.
61. *Committee on Bill*, 695.
62. *Times*, 3 June 1830, 3b. Also see report in *The News*, 6 June 1830, 4a–b.
63. A. Friedman, 'Goldsmith and the Weekly Magazine', *Modern Philology*, 1935, 32: 281–99. Goldsmith was a physician.
64. C. Bruce, *Mirth and Morality: A Collection of Original Tales*, London: T. Tegg and Son, 1834, 29–40, 39.
65. Ibid., 256.
66. *Times*, 3 August 1824, 3b.
67. Ibid.
68. *Times*, 3 August 1824 3b.
69. Ibid., 18 May 1830, 2b.
70. Ibid., 15 August 1830, 5c.
71. Ibid., 13 July 1825, 3b.
72. Ibid.
73. H. Ritvo, *The Animal Estate: The English and Other Creatures in the Victorian Age*, Cambridge, MA: Harvard University Press, 1987, 85–93.
74. *Annals of Sporting and Fancy Gazette*, 1822, 2: 5.
75. *Times*, 13 July 1825, 3b.
76. Ibid., 10 June 1830, 2f.
77. *Voice of Humanity, The Voice of Humanity for the Communication and discussion of al subjects relative to the conduct of Man and the inferior Animal Creation*, 1830, 1: 30.
78. HO44/20 Letter from Scotics to Robert Peel, 2 June 1830.
79. Anon, 'Remarks on Hydrophobia and its effects on the Dog and the Human Race', *Annals of Sporting and Fancy Gazette*, 1822, 2: 5.
80. *Voice of Humanity*, 1830, 1: 30.
81. *Times*, 5 June 1830, 6d.
82. 'Indictment against the Westminster Pit', *Voice of Humanity*, 1830, 1: 24–25.
83. Ibid.
84. William Hamilton Drummond, 'Humanity to Animals the Christian Duty', *Voice of Humanity*, 1830, 1: 58.
85. 'Indictment against the Westminster Pit', 24–25.
86. HO44/14. Anonymous letter to Robert Peel, 30 June 1830.
87. F. M. Thompson, *Voice of Humanity*, 1830, 1: 37.
88. J. D. Blaisdell, 'An Ounce of Prevention Causes a Ton of Concern: Rabies and the English Dog Tax of 1796' *Veterinary History*, 2000–2001, 10: 129–46.
89. Youatt, 'On Rabies Canina', 87.

90. HO 44/18. Letter from Henry Thompson to Sir Robert Peel, 1828.
91. J. Wade, *Colquhoun's A Treatise on the Police and Crimes of the Metropolis*, London: Longman and Co., 1829, 351.
92. 'A Constant Reader', *Times*, 4 June 1830, 3c.
93. M. D. George, *A Catalogue of Political and Personal Satires in the British Museum*, Vol. 11, London: British Museum, 1952, [16147], 163. Dogberry was the chief of police in Shakespeare's 'Much Ado about Nothing' and a figure of comic incompetence. Wood enjoyed a not dissimilar reputation – 'ignorant and self-satisfied' – which proved a handicap in his attempts to persuade the Commons to take dog control legislation seriously in 1830 and 1831.
94. Ibid., [16234], 336.
95. Ibid., [16153], 403.
96. HO 44/20, 363. Letter from Anon to Sir Robert Peel, June 1830.
97. HO 44/20, 361. Letter from Anon to Sir Robert Peel, June 1830. *Morning Advertiser*, 2 June 1830, 3a.
98. Ibid.
99. Committee on Bill, 695–97. The Committee was chaired by Alderman Wood, MP for the City of London.
100. Veterinarians were Edward Coleman and William Youatt. Medical witnesses included William Babbington, Benjamin Collins Brodie, Henry Earle, Richard Frankam, John Morgan, Alexander Thomson, Anthony Todd Thompson, and Benjamin Travers. Magistrates included Lancelot Baugh Allen and Joseph Terry Home; Edward Thomas Handley represented the police. John Ody provided an account of dog life on the Strand.
101. Coleman was head of the college from 1793 to 1839, and was almost exclusively interested in horses.
102. *Committee on Bill*, 715.
103. Ibid., 692.
104. Ibid., 694.
105. 'The absolute number of people who die of hydrophobia, as compared with the mass of society is small; but the chief thing from which society requires to be delivered is the alarm which they are under on the subject' Committee on Bill, 710.
106. HO 44/20, 361–62. Letter from Anon to Sir Robert Peel, June 1830.
107. B. Harrison, 'Animals and the State in Nineteenth century England', *English Historical Review*, 1973, 88: 788–89.
108. W. Youatt, 'Introductory Lecture on Veterinary Medicine and Surgery', *Lancet*, 1831, ii: 78.
109. Ibid., 80.
110. Murray, *Remarks on the Disease*, vii.
111. Bardsley, 'Hydrophobia', 483.
112. Anon, 'Remarks on the Disease Called Hydrophobia, Prophylactic and Curative', *Westminster Review*, Oct 1830, 419.

2   Rabies at Bay: 'The Dog Days', 1831–1863

1. On zymotic diseases see M. Worboys, *Spreading Germs: Disease Theories and Medical Practice in Britain, 1865–1900*, Cambridge: Cambridge University Press, 2000, 34–42.

2. C. Hamlin, *Public Health and Social Justice in the Age of Chadwick: Britain, 1800–54*, Cambridge: Cambridge University Press, 1998.
3. Words taken from titles of articles on cases on hydrophobia reported in the *Times* between 1 January 1831 and 31 December 1859.
4. *Animal's Friend, or the Progress of Humanity*, 1840, 8: 72.
5. Ibid., 1841, 9: 20–1.
6. 'Death of Sir Mathew Wood', *Times*, 27 September 1843, 5b–c. Wood seems to have had few friends; his obituarist in the *Times* marvelled that someone with so little talent had risen to such high rank, though this may have said more about the writer's prejudices about someone with a background in trade and who championed unpopular causes.
7. *Times*, 10 June 1846, 6b.
8. G. Cruikshank, *The Comic Almanack for 1836*, London: Charles Tilt, 1836.
9. *Cruikshank's Comic Almanack for 1837*, London: Charles Tilt, 1837, 96.
10. *Cruikshank's Comic Almanack for 1853*, London: David Bogue, 1853, 397. Thames water was notoriously polluted, as evident in the Great Stink of 1858, and understood to be a major source of disease.
11. The Association for Promoting Rational Humanity towards the Animal Creation proposed two Bills in 1831 to control conditions in cattle markets. *House of Lords Journal*, 1830–31, 63: 993 and 1831–32, 64: 192.
12. A. F. Pollard, and Charlotte Fell-Smith, 'Pease, Edward (1767–1858)', rev. M. W. Kirby, Oxford *Dictionary of National Biography*, Oxford University Press, 2004; online edn, Oct 2006 [http://www.oxforddnb.com/view/article/21728, accessed 18 Dec 2006].
13. *Select Committee on Bill to Consolidate Laws Relating to Improper Treatment of Animals. Report, Minutes of Evidence*, Parl. Papers, 1831–1832 (67), v, 73.
14. Ibid., 101.
15. C. Otter, 'Cleansing and Clarifying: Technology and Perception in Nineteenth Century London', *Journal of British Studies*, 2004, 43: 40–64.
16. *Select Committee on Bill*, 91.
17. Smithfield and its livestock market was already a place of notoriety for cruelty to animals. See D. Donald, 'Beastly Sights': The Treatment of Animals as a Moral Theme in Representations of London c. 1820–50', *Art History*, 1999, 22: 514–44.
18. *Select Committee on Bill*, 91.
19. B. Harrison, 'Animals and the State in Nineteenth Century England', *English Historical Review*, 1973, 88: 786–820.
20. *Times*, 3 April 1846, 8c.
21. Ibid.
22. *Times*, 14 December 1837, 3a.
23. *SPCA, Annual Report for 1835*, 29.
24. W. Schivelbusch, *The Railway Journey: The Industrialization of Time and Space in the Nineteenth Century*, Berkley: University of California Press, 1986.
25. *SPCA Annual Report for 1836*, 69–70.
26. *Times*, 30 August 1838, 5c–e.
27. *Voice of Humanity*, 1830, 1: 30. Letter from Maria Thompson.
28. Dog muzzles seemed to some reformers to resemble the 'unnecessary' harness customarily used on beasts of burden. It infringed on a dog's capacity to discharge heat, arousing fever in the dog.

29. 'How to Make a Dog Mad', *Cruikshank's* ... *1837*, 96.
30. *Select Committee on Metropolis Police Offices, Report, Minutes of Evidence, Appendix*, Parl. Papers, 1837–1838, (578), xv, 365.
31. *Bill for Further Improving Police in and Near Metropolis*, Parl. Papers, 1839, (58) iv, 409–41.
32. *Select Committee on Metropolis Police Offices*, 321–622.
33. *SPCA Annual Report*, 1836, 70.
34. See for example Manchester in 1836. Manchester Archives, *Annual Report of the Lamp, Scavenging, Fire Engine, and Nuisances Committee*, 1836, 191, 196, 322 and 456.
35. Watch Committee for the Borough of Manchester, *Criminal and Miscellaneous Returns of the Manchester Police for the Year 1846*.
36. *Manchester Guardian*, 16 May 1846, 10e–f.
37. *Annual Report of the Lamp, Scavenging*, 454.
38. *Hansard*, 184, 17 July 1866, 929.
39. *Bill to Amend Acts for More Effectual Prevention of Cruelty to Animals*, Parl. Papers, 1854, (106), ii, 145–49. See M. Radford, *Animal Welfare Law in Britain: Regulation and Responsibility*, Oxford: Oxford University Press, 2001, 63–65 and 82.
40. *Hansard*, 134, 26 June 1854, 639; 9 July 1864, 1075–80; 10 July 1854, 1429–36.
41. *Times*, 13 July 1854, 10d. Also see 10 July 1854, 10c. had the argument of Sir Grantley F. Berkeley in the early 1840s. *Times*, 31 October 1846, 5a.
42. *Times*, 20 March 1843, 7a.
43. Ibid.
44. *Times*, 17 March 1843, 5a.
45. *Times*, 9 May 1853, 6b. Also see *Times*, 2 November 1852, 3e.
46. *Times*, 27 July 1854, 12d.
47. *Times*, 6 April 1853, 8b.
48. *Times*, 27 July 1854, 12d.
49. This was directly linked by one correspondent to raising revenues for the Crimean War. See *Times*, 27 July 1854, 12d.
50. *Times*, 23 July 1855, 12f.
51. Discussions of the Brontës and dogs to date have ignored rabies.: I. Kreilkamp, 'Petted Things: Wuthering Heights and the Animal', *Yale Journal of Criticism*, 2005, 18: 87–110; M. B. Adams, 'Emily Brontë and Dogs: Transformation Within the Human-Dog Bond', *Society and Animals*, 2000, 8: 167–81.
52. Emily Brontë, *Wuthering Heights*, Harmondsworth: Penguin, 2003; Charlotte Brontë, *Shirley*, Harmondsworth: Penguin, 1985.
53. E. Brontë, *Wuthering Heights*, 15.
54. Ibid., 17.
55. Ibid.
56. Ibid., 75. Cropping was a common practice of trimming a dog's ears to make it look fiercer, or in fighting dogs to make its ears a smaller target to attack.
57. Ibid., 49.
58. Ibid., 53.
59. Ibid., 158.
60. Ibid., 162.

61. On the significance of fasting see S. Gubar, 'The Genesis of Hunger, According to *Shirley'*, *Feminist Studies*, 1976, 3: 16–17.
62. C. Bronte, *Shirley*, 476.
63. Ibid., 478.
64. Ibid., 479.
65. J. L. Bardsley, 'Hydrophobia', in J. Forbes, A. Tweedie and J. Conolly, eds, *The Cyclopaedia of Practical Medicine*, London: Sherwood, Gilbert and Piper, 1833, 483–518.
66. Ibid., 518.
67. Pettigrew, *Hydrophobia*, MS, Wellcome Library, London.
68. W. Youatt, 'Remarks on the Case of Supposed Recovery from Hydrophobia', *Veterinarian*, 1835, 8: 508–11, in *Lancet*, 1835, 24: 714–17. In 1837 he wrote on cases of dogs recovering from rabies: W. Youatt, 'Preventive and Curative Treatment of Rabies', *Veterinarian*, 1837, 10: 325, 389, 445, and 517, reprinted in the *Lancet*, 1837, i: 55–8.
69. J. Elliotson, 'Lectures on the Theory and Practice of Medicine', *London Medical Gazette*, 1833, 500–1.
70. T. Tomkins and E. S. Varenne, 'Recovery in a Case of Hydrophobia', *Lancet*, 1835, 24: 630–31; J. L. M'Carthy, 'Case Frequently Simulating Hydrophobia', *Lancet*, 1835, 25: 25–6.
71. F. Eagle, 'On the Identity in Nature, Symptoms, Morbid Appearances and Causes, of Hysteria and Hydrophobia', *Lancet*, 1836, 28: 583–89.
72. Ibid., 588., there were a number of reports of hydrophobia developing after the bite of a non-rabid dog. Were these also cases of the spurious disease, or were these instances where a dog incubating the disease was able to transmit the virus? In the case of the latter, could any dog bite be regarded as safe? E. L Knowles, 'Hydrophobia', *Lancet*, 1834, 22: 928–29.
73. W. Youatt, 'Origin of Hydrophobia', *Lancet*, 1839, 32: 109. Youatt was typical in writing as though rabies and hydrophobia were a single disease, though it was increasingly common to distinguish between the animals and human forms.
74. Paradoxically, in the 1848–1849 cholera outbreak in Bradford dogs were suspected to be carriers of the disease. They were ordered to be confined and strays were killed. M. Sigsworth, *Cholera in the West and East Ridings of Yorkshire, 1848–1892*, Unpublished DPhil Thesis, Sheffield City Polytechnic, 1991.
75. 'Remarks on Hydrophobia', *Lancet*, 1852, 59: 453–54. see *British and Foreign Medico-Chirurgical Review*, 1852, 10: 539–41.
76. Ibid., 453.
77. T. L. Kemp, 'Rabies and Hydrophobia', *Edinburgh Medical and Surgical Journal*, 1855, 58: 1–19.
78. Ibid., 12.
79. J. Netten Radcliffe, 'Materials towards the formation of a better knowledge of hydrophobia', *Lancet*, 1855, i: 153–4 and 258–60; 1856, i: 683–5; 1856, ii: 217 and 271–3.
80. W. Youatt, 'Animal Pathology: Lecture XXI', *Veterinarian*, 1838, 11, 232–41, 276–89, 388–98, and 456–67.
81. Pritchard, 'Rabies in a Cow and Pigs, and on Rabies Generally', *Veterinarian*, 1836, 9: 199–201.
82. J. Secord, *Victorian Sensation: The Extraordinary Publication, Reception, and Secret Authorship of Vestiges of the Natural History of Creation*, Chicago: Chicago University Press, 48–51.

83. E. Phillips, 'Dick, William (1793–1866)', *Oxford Dictionary of National Biography*, Oxford University Press, 2004 [http://www.oxforddnb.com/view/article/56035, accessed 12 Jan 2007].

84. Quoted in the *Veterinarian*, 1841, 14: 23.

85. *Veterinarian*, 1844, 17: 645.

86. W. C. L. Martin, *The History of the Dog: Its Origin, Physical and Moral Characteristics, and its Principal Varieties*, London: Charles Knight & Co., 1845; H. D. Richardson, *The Dog; Its Origin, Natural History and Varieties. With Directions for Its General Management*, London: Wm. S. Orr & Co., 1851; E. Mayhew, *Dogs: Their Management. Being a New Plan of Treating the Animal, Based upon a Consideration of His Natural Temperament*, London: George Routledge and Co., 1854.

87. F. Galton, 'Heredity Talent and Character', *Macmillan's Magazine*, 1865, 12: 157–66 and 318–27.

88. Edward Mayhew was a veterinarian who had edited Blaine's *Veterinary Art*, hence, it is unsurprising that his volume contained long sections on disease.

89. Mayhew, *Dogs*, 157.

90. Ibid., 163–64.

91. Editorial, *Veterinarian*, 1852, 26: 472–75.

92. Worboys, *Spreading Germs*, 108–49.

93. *Veterinarian*, 1856, 29: 341–43.

94. Ibid., 343.

95. *Veterinarian*, 1848, 21: 278.

96. Ibid., 586.

97. Stonehenge [John Henry Walsh], *The Dog, in Health and Disease*, London: Longmans, Green, and Co, 1859.

98. Ibid., 386.

99. Ibid., 337.

100. G. H. Lewes, 'Mad Dogs', *Blackwood's Magazine*, 1861, 90: 222–40.

101. Rosemary Ashton, 'Lewes, George Henry (1817–1878)', *Oxford Dictionary of National Biography*, Oxford University Press, 2004 [http://www.oxforddnb.com/view/article/16562, accessed 16 Nov 2006].

102. D. Noble, 'George Henry Lewes, George Eliot and the Physiological Society', *Journal of Physiology*, 1976, 263(1): 45–54.

103. Quoted in J. B. West, 'Spontaneous Combustion, Dickens, Lewes, and Lavoisier', *Physiology*, 1994, 9: 276.

104. Lewes, 'Mad Dogs', 222–24.

105. G. Cottesloe, *Lost, Stolen or Strayed. The Story of the Battersea Dogs' Home*, London: Arthur Barker, 1971.

106. H. F. 'The Home for Lost and Starving Dogs', *The Field*, 1860, 16: 434.

107. *Times*, 18 October 1860, 6e.

3   Rabies Resurgent: 'The Dog Plague', 1864–1879

1. Ernest Clarke, 'Fleming, George (1833–1901)', rev. Linda Warden, *Oxford Dictionary of National Biography*, Oxford University Press, 2004 [http://www.oxforddnb.com/view/article/33167, accessed 20 Oct 2006].

2. J. K. Walton, 'Mad Dogs and Englishmen: The Conflict Over Rabies in Late Victorian England', *Social History*, , 13: 219–39.

3. 'The Two Dog Shows', *Liverpool Daily Post*, 4 June 1864, 4c.
4. 'The Mad Dog Crusade', *The Liverpool Daily Post*, 27 May 1864, 9c; 'Any Cause for Alarm?', *Liverpool Daily Post*, 31 May 1864, 9c.
5. 'The Mad Dog Panic', *Liverpool Daily Post*, 1 June 1864, 2c.
6. The editorial asked, 'who shall decide when doctors disagree?' 'Any Cause for Alarm?' *Liverpool Daily Post*, 31 May 1864, 9c. As another correspondent put it, 'Has medical science yet attained to anything but what is merely empirical on the subject of canine madness?', 'A Plea for the Dogs', *Liverpool Daily Post*, 31 May 1864, 10e.
7. *Liverpool Mercury*, 2 June 1864, 4 c–d.
8. 'There was no recorded example of a female case of hydrophobia, and that what, then, is less likely to produce the effects of quietness and calmness than on the one hand placarding the walls with proclamations which alarm timid people and on the other, compelling the owners of dogs to fasten them up', 'Any Cause for Alarm?' *Liverpool Daily Post*, 31 May 1864, 9c.
9. *Liverpool Mercury*, 11 June 1864, 6e.
10. *Liverpool Daily Post*, 3 June 1864, 3d.
11. 'The Two Dog Shows' *Liverpool Daily Post*, 4 June 1864, 4c.
12. One correspondent, who signed himself 'Sufferer', explained, 'to be unnecessarily summed before a magistrate is sufficiently annoying and inconvenient to a man of business, without the addition of insolence from the informer'. *Liverpool Daily Post*, 23 May 1864, 3d.
13. *Liverpool Daily Post*, 15 June 1864, 7f.
14. *Liverpool Mercury*, 15 June 1864, 8d.
15. *Liverpool Daily Post*, 23 May 1864, 7d.
16. *Liverpool Daily Courier*, 1 June 1864, 5b.
17. 'A Dogs' Home for Liverpool', *Liverpool Daily Post*, 31 May 1864, 10e; *Daily Courier*, 1 June 1864, 5a.
18. *Liverpool Mercury*, 18 May 1864, 5d.
19. *Liverpool Mercury*, 19 May 1864, 5e.
20. *Manchester Guardian*, 1 June 1864, 2g; *Stockport Advertiser*, 3 June 1864, 3f.
21. *Stockport Advertiser*, 27 May 1864, 2g.
22. In the words of one bewildered 'dog fancier' who spoke out about the folly of the measures, 'I will not attempt to discuss the probability of dogs getting mad in this mild and temperate season. It is true we have had a few hot days; but I have yet to learn that in Manchester it had produced any symptoms of hydrophobia among the canine species'. *Manchester Courier*, 15 June 1864, 4f.
23. J. Ridley, *Lord Palmerston*, London: Constable, 1970, 579.
24. J. Morley, *The Life of William Ewart Gladstone*, London: Macmillan and Co., 1905, 780.
25. F. M. Turner, 'Rainfall, Plague and the Prince of Wales: A Chapter in the Conflict of Religion and Science', *Journal of British Studies*, 1974, 13: 46–65.
26. M. Worboys, *Spreading Germs: Disease Theories and Medical Practice in Britain, 1865–1900*, Cambridge: Cambridge University Press, 2000, 58 and 62.
27. G. H. Lewes, 'Mad Dogs', *Blackwood's Magazine*, 1861, 90: 222–40.
28. *Stockport Advertiser*, 11 May 1866, 6d.
29. *Times*, 20 February 1866, 12g.
30. *Times*, 9 March 1866, 11g.
31. *Times*, 8 May 1866, 11g.

32. *Times*, 26 June 1865, 12e.
33. *Manchester Courier*, 17 April 1866, 6f.
34. 'Dog Slaughter', *Manchester Guardian*, 24 April 1866, 6f.
35. *Manchester City News*, 2 June 1866, 3f.
36. *Hansard*, 184, 17 July 1866, 929.
37. At the same time, there was also a case for lowering the tax by bringing dog ownership within the scope of the 'respectable working-man', and that the fear of rabies should not infringe the rights of the respectable poor. This was also the view of William Gladstone who explained that, 'the very persons what the tax is designed to reach cannot afford to pay so much' and they 'reconcile it to their consciences to pay nothing at all.' *Hansard*, 181, 9 March 1866, 1834–37.
38. Ibid.
39. Ibid.
40. *A Bill to Repeal the Duties of Assessed Tax on Dogs and Impose in Lieu Thereof a Duty of Excise*, Parl. Papers, 1867, (36), ii, 609.
41. *Times*, 16 March 1867, 12c.
42. *Times*, 12 August 1867, 10c.
43. *Lancet*, 1867, i: 687.
44. *Times*, 20 August 1867, 5e.
45. J. Winter, *London's Teeming Streets*, London: Routledge, 1993, 47.
46. *Traffic Regulation Bill*, 678–79.
47. *Clerkenwell News and London Times*, July 10 1868, 6e. Winter has noted that 'it is ironic that [Mayne] had to weather a storm of abuse in the last year of his long career for supposedly becoming obsessed with muzzling dogs ... when, in fact, he heartily disliked involving his men in such assignments.' Winter, *London's*, 47.
48. *Daily News*, 2 July 1868, 5b–c.
49. *Penny Illustrated Paper*, 11 July 1868, 18.
50. Quoted in C. Warren, 'Dogs in London', *Contemporary Review*, 1887, 51: 109–10. Warren was Chief Commissioner of the Metropolitan Police between 1886 and 1888. Keith Surridge, 'Warren, Sir Charles (1840–1927)', *Oxford Dictionary of National Biography*, Oxford University Press, Sept 2004; online edn, May 2006 [http://www.oxforddnb.com/view/article/36753, accessed 19 Dec 2006].
51. 'Legalised Cruelty', *Lancet*, 1868, i: 16.
52. *Times*, 21 September 1868, 11f.
53. For a history of the Home see Gloria Cottesloe, *The Story of the Battersea Dogs' Home*, London: David & Charles, 1979. Also see Hilda Kean, *Animal Rights: Political and Social Change in Britain since 1800*, London: Reaktion, 1998, especially chapter three 'Continuity and Change: Fallen Dogs and Victorian Tales'.
54. 'The Fate of the Dogs', *Morning Post*, 1 July 1868, 6b.
55. *Return of the Number of Deaths from Hydrophobia In England during each of the Nine Years from 1866–1875*, Parl. Papers, 1877 (163), xlix, 251.
56. *Halifax Guardian*, 30 January 1869, 2c.
57. *Preston Guardian*, 2 December 1868, 3c; 13 February 1869, 3d.
58. *Preston Guardian*, 27 March 1869, 4b.
59. *Return of the Number of Deaths from Hydrophobia in England 1866–1877*, Parl. Papers, 1878, (120), xlvi, 158–59.

60. *The Field*, 20 May 1871, 413.

61. Ibid.

62. *Standard*, 16 June 1871. This article is in a collection of miscellaneous pamphlets and articles of George R. Jesse held at the British Library. See George Richard Jesse, *A collection of cuttings, etc., relating to Hydrophobia collected by G. R. Jesse*, London: British Library, 1873–79.

63. *Association for the Protection of Dogs and the Prevention of Rabies*, Pamphlet, 1878, Jesse, *A collection of cuttings, etc.*

64. Ibid.

65. 'Beth Gelert' first wrote to the *Macclesfield Guardian*, but the letter was inserted in the *Times* on 29 October 1874 at a cost of £15 to Jesse, see letter in *A collection of cuttings, etc.* The 1811 poem was based on a legend which tells the story of the hound of the thirteenth century Prince Llywelyn which runs to his master covered in blood and is assumed to have killed the Prince's son. It turns out the Prince's son is alive and that Beth Gelert was blood-stained because he had killed a large wolf in the child's bedroom. Allegedly, the Prince was so full of remorse that he never smiled again. The resonance with claims of many non-rabid dogs being slain is clear.

66. J. Ruskin, *Fors Clavigera: Letters to the Workmen and Labourers of Great Britain*, Vol. 4, Orpington: George Allen, 1874.

67. Jesse, *A collection of cuttings, etc.*

68. When the narrator first approaches the Chowne's rectory, the presence of dogs is deployed to intensify the viciousness that characterises the dwelling that was the home of the Parson. 'A tremendous roaring of dog broke upon me the moment I got the first glimpse of the house ... One huge fellow rushed up to me, and disturbed my mind to so great a degree that I was unable to take heed of anything about the place except his savage eyes and highly alarming expression and manner. For he kept on showing his horrible tusks and growling a deep growl broken with snarls, and sidling to and fro, so as to get the better chance of a dash at me' R. D. Blackmore, *The Maid of Sker*, London: Sampson Low, Marston & Co., 1872, 175.

69. Ibid., 464.

70. Ibid., 464.

71. Ibid., 467.

72. Parson Jack was a changed man, 'His wife found the benefit of this change, and so did his growing family, and so did the people who flocked to see his church, in the pleasure of being afraid of him. In the roads, he might bite, but in his surplice, he was bound to behave himself, or at least, he must bite the churchwarden first. Yet no one would have him sprinkle a child, until a whole year was over'. So profoundly moved was he by his brush with hydrophobia that he was able to preach a sermon that 'stirred every heart', with the text, 'Is thy servant a dog, that he should do this thing?' Ibid., 469.

73. *Manchester Guardian*, 17 January 1874, 9d.

74. *Manchester Guardian*, 31 August 1874, 7b.

75. *Animal World*, 1874, 5: 163–65. Reilly contrasted his help for the suffering people of England, against the attitude of the government that had recently offered £1,000 for a cure for snake bites in India.

76. *Manchester Guardian*, 23 January 1874, 7c.

77. *Manchester Guardian*, 1 May 1874, 7c.

78. G. Fleming, *A Manual of Veterinary Sanitary Science and Police*, London: Chapman and Hall, 1875.
79. The case was put most forcefully and unsurprisingly by John Gamgee. See *Times*, 1 May 1874, 11e. On Gamgee's life and work see Sherwin A. Hall, 'Gamgee, John (1831–1894)', *Oxford Dictionary of National Biography*, Oxford University Press, 2004 [http://www.oxforddnb.com/view/article/56034, accessed 17 Jan 2007].
80. Charles Kent, 'Berkeley, (George Charles) Grantley Fitzhardinge (1800–1881)', rev. Julian Lock, *Oxford Dictionary of National Biography*, Oxford University Press, 2004[http://www.oxforddnb.com/view/article/2212, accessed 19 Dec 2006].
81. The term 'the fancy' was coined for those who fancied the pursuit of certain sports, like hunting and coursing.
82. J. B. Sanderson, *Times*, 18 May 1874, 9d. T. Romano, *Making Medicine Scientific: John Burdon Sanderson and the Culture of Victorian Science*, Baltimore: Johns Hopkins University Press, 2002.
83. G. Fleming, *Rabies and Hydrophobia: Their History, Nature, Causes, Symptoms and Prevention*, London: Chapman and Hall, 1872, 192–93.
84. *Times*, 18 May 1874, 9d.
85. Fleming, *Rabies*, 195.
86. Fleming contrasted the domestic disposition of dogs with those of cats, stating that felines leave 'the domestic roof when rabid, though rather through the influence of its savage nature than its devotion to its human companions.' Ibid., 248.
87. Ibid., 2.
88. Ibid., 4–6. Also see *Animal World*, 1875, 6: 178–79.
89. Ibid., 351.
90. In 1886–1887 great play was made by antivivisectionists of Fleming having changed his mind to reject spontaneous generation altogether, see Ch 4.
91. Fleming, *Rabies*, 124.
92. Worboys, *Spreading Germs*, 58 and 62.
93. Fleming suggested that the virus could become panzootic in many species, hence, 'Contaminated animals may live long enough to inoculate several others, and some of these may again become active agents in the propagation of the poison, and wander into different localities'. Fleming, *Rabies*, 91.
94. *Times*, 19 October 1872, 7b.
95. Fleming, *Rabies*, 353.
96. *Animal World*, 1878, 9: 119–20.
97. Turner, 'Rainfall, Plague', 46–65.
98. Hon. Grantley F. Berkeley, *Facts Against Fiction: The Habits and Treatment of Animals Practically Considered; Hydrophobia and Distemper; with Some Remarks on Darwin*, London: Samuel Tinsley, 1874.
99. Ibid., 135.
100. Ibid., 'ladies and their waiting maids have been bitten by their canine pets, and when there has been any constitutional or nervous tendency in the mind – a species of morbid apprehension, needless in its origin, but very difficult to combat – has sprung up, capable by its own chimeric and mental poison to produce, not the very disease that caused the unfounded dread, but a madness from the force of imagination, arising in and mastering the temporarily unhinged intellect, and which might ultimately produce death'. 344.

101. J. L. Brand, *Doctors and the State: The British Medical Profession and Government Action in Pubic Health, 1870–1912*, Baltimore, MA: The Johns Hopkins University Press, 1965, 22–84.
102. *Times*, 18 May 1874, 9d.
103. *Lancet*, 1874, i: 734–35 and ii: 655.
104. *Lancet*, 1877, ii: 136.
105. *Times*, 3 January 1877, 3c.
106. Worboys, *Spreading Germs*, 8–12.
107. *Registrar General of Births, Deaths and Marriages in England Fortieth Annual Report for 1876*, Parl. Papers, 1877 [C.2075], xxii, 343.
108. L. Lindsay, *Mind in the Lower Animals*, London: Kegan Paul & Co., 1879, 363.
109. *Registrar General of Births*, [C.2075], 290–95.
110. *Registrar General of Births, Deaths and Marriages in England Fortieth Annual Report*, Parl. Papers, 1878–1879 [C.2276], xix, 284–85.
111. Ibid., 399.
112. *Lancet*, 1877, i: 63.
113. *Lancet*, 1877, ii; 399 and 660. The writer observed that 'the Veterinary Department of the Privy Council does not recognise the existence of rabies', being critical of officials' exclusive focus on diseases named in the Contagious Diseases (Animals) Act.
114. *Times*, 31 October 1877, 9d.
115. *Times*, 6 November 1877, 4f.
116. The editorial recommended that when assailed by a large dog, it was best to avoid using one's hands and instead that a man should 'take off his hat and use it as sort of shield'. Ibid. Also see *Times*, 2 November 1877, 6f.
117. *Animal World*, 1877, 8: 178–79.
118. *Times*, 23 November 1877, 4a. Also see G. Cottesloe, *Story of the Battersea Dogs' Home*, Newton Abbot: David & Charles, 1979. The number of dogs collected and the noise they made stimulated the opposition of the neighbours in Holloway and its move to Battersea.
119. *Annual Report for the Temporary Home of Lost and Starving Dogs for 1875*, 7.
120. *Annual Report for the Temporary Home of Lost and Starving Dogs for 1878*, 7.
121. *Times*, 23 November 1877, 4a.
122. *Pall Mall Gazette*, 5 November 1877, 10–11.
123. *Animal World*, 1 December 1877, 148.
124. J. E. Strick, *Sparks of life: Darwinism and the Victorian Debates Over Spontaneous Generation*, London: Harvard University Press, 2000; Worboys, *Spreading Germs*, 62, 86–90 and 152–54.
125. W. Lauder Lindsay, 'The Artificial Production of Human Diseases in Lower Animals', *Lancet*, 1878, i: 380–81.
126. Worboys, *Spreading Germs*, 43–72 and 73–107.
127. Watson was best known as the author of Britain's most popular medical textbook of the century. Norman Moore, 'Watson, Sir Thomas, First Baronet (1792–1882)', rev. Anita McConnell, *Oxford Dictionary of National Biography*, Oxford University Press, 2004 [http://www.oxforddnb.com/view/article/28871, accessed 19 Dec 2006].
128. T. Watson, 'The Abolition of Zymotic Diseases' *Nineteenth Century*, 1877, 1: 380–96.
129. *Lancet*, 1877, i: 655–56. Also see Worboys, *Spreading Germs*, 147.
130. T. Watson, 'Hydrophobia and Rabies', *Nineteenth Century*, 1877, 2: 736, 717–36.

131. T. M. Dolan, *The Nature and Treatment of Rabies or Hydrophobia. Being the Report of the Special Commission Appointed by the Medical Press and Circular, with Valuable Additions*, London: Baillière, Tindall, and Cox, 1878.
132. This division of labour was given by Dolan in evidence to the *Select Committee of House of Lords on Rabies in Dogs. Report, Proceedings, Minutes of Evidence, Appendix*, Parl. Papers, 1887, (322), xi, 451.
133. Dolan, *The Nature*, 16.
134. Ibid.
135. Ibid., 119.
136. *Lancet*, 1878, ii: 329–31.
137. *Lancet*, 1880, i: 755–57.
138. J. Fayrer, 'Hydrophobia', *Lancet*, 1877, ii: 785. The case was written up in: Idem, *Clinical and Pathological Observations in India*, 1873, 545–51. We have been unable to find the article mentioned in *Blackwood's Magazine* or any other nineteenth-century periodical.
139. For an introduction to degenerationist ideas see A. T. Scull, C. MacKenzie and N. Hervey, *Masters of Bedlam*, Princeton: Princeton University Press, 1996.
140. M. Micale, *Approaching Hysteria: Disease and Its Interpretations*, Princeton: Princeton University Press, 1995.
141. D. Tuke, *Illustrations of the Influence of the Mind upon the Body in Health and Disease*, London: J & A Churchill, 1872.
142. D. Trotter, 'The Invention of Agoraphobia', *Victorian Literature and Culture*, 2004, 32: 463–74.
143. P. Errera, 'Some Historical Aspects of the Concept, Phobia', *Psychiatric Quarterly*, 1962, 36: 332–34.
144. Tuke, *Illustrations of the Influence*, 66.
145. Ibid., 210–12.
146. W. Lauder Lindsay., 'Spurious Hydrophobia in Man', *Journal of Mental Science*, 1878, 26: 549–59; continued in *Journal of Mental Science*, 1879, 27: 51–64.
147. S. Barrows, *Distorting Mirrors: Visions of the Crowd in Late Nineteenth century France*, London: Yale University Press, 1981.
148. Lindsay, 'Spurious Hydrophobia', 61. Also see L. Lindsay, *Mind in the Lower Animals in Health and Disease*, Vol. II, London: C Kegan Paul & Co., 1879, 175–78 and 363.
149. Ibid., 55.
150. G. Berrios and R. Porter, *The History of Clinical Psychiatry*, London: Athlone, 1995, 546; G. E. Berrios, *The History of Mental Symptoms*, Cambridge: Cambridge University Press, 1996, 147–78 and 263–70.
151. Berrios, *The History of Mental Symptoms*, 270–71.
152. *Times*, 31 October 1877, 10e; *Lancet*, 1877, ii: 810–11.
153. Henry W. T. Ellis reported on Chloral: 'friends of the poor little sufferer fully appreciated the beneficial effects of the remedy used'. *Lancet*, 1871, ii: 112.
154. *British Medical Journal (BMJ)*, 1877, ii: 862.
155. R. D. French, *Antivivisection and Medical Science in Victorian England*, Princeton: Princeton University Press, 1975.
156. *Animal World*, 1877, 8: 178–79.
157. *BMJ*, 1877, ii: 824.

158. Ibid.
159. Watson, 'Hydrophobia', 717–36.
160. Fleming, *A Manual of Veterinary Sanitary*, II, 264.
161. Ibid., 269.
162. *Return of Number of Deaths from Hydrophobia in England and Wales, 1866–77; Return of Number of Dogs Assessed to Dog Tax, 1866–68; Return of Number of Licences Granted for Dogs, 1868–77*, Parl. Papers, 1878 (120), xlvi, 157.
163. *Hansard*, 239, 15 April 1878, 1324.
164. Ibid., 1320–51. Also see *Times*, 12 April 1878, 6b and 16 April 1878, 7b.
165. Mr Assheton, (Clitheroe), Ibid., 1326.
166. Ibid., 1331–37. This was also the view of George Goschen (City of London), 1326–27.
167. Ibid., Mr Hopwood (Stockport) stated the tax 'would seal the death warrant of many a humble favourite throughout the kingdom'.
168. Ibid., 1350.
169. R. Caldecott, *The Mad Dog*, London: Frederick Warne and Co. Ltd, 1879.

4 Rabies Cured: 'The Millennium of Pasteurism', 1880–1902

1. G. L. Geison, 'Louis Pasteur', in C. C. Gillespie, ed., *The Dictionary of Scientific Biography*, New York: Charles Scribner, 1974, 350–416; G. L. Geison, *Private Science of Louis Pasteur*, Princeton: Princeton University Press, 1995; M. D. Reynolds *How Pasteur Changed History: The Story of Louis Pasteur and the Pasteur Institute*, Bradenton, Fla.: McGuinn & McGuire Pub., 1994.
2. B. Hansen, 'America's First Medical Breakthrough: How Popular Excitement About a French Rabies Cure in 1885 Raised New Expectations for Medical Progress', *American Historical Review*, 1998, 103: 373–418.
3. *Zoophilist*, 1883–84, 3: 262. Little has been written about antivivisectionist activity in the 1880s as a whole, though there is some comment in R. D. French, *Antivivisection and Medical Science in Victorian England*, Princeton: Princeton University Press, 1975.
4. An indication of the continuing uncertainties and anxieties is that in 1879, the Royal College of Physicians had announced a £100 prize, sponsored by an *Medical Press*, for the best essay on 'Hydrophobia: Its Nature, Prevention and Treatment'. *Lancet*, 1880, i: 305. For examples of reports on hydrophobia in 1881 see *Times*, 26 January 1881, 13g; 22 August 1881, 6c; 15 October 1881, 9f; 3 November 1881, 3e.
5. *British Medical Journal (BMJ)*, 1881, ii: 807–11; 1884, i: 186.
6. R. Neale, 'Suggestions for the Treatment of Hydrophobia', *Lancet*, 1881, ii: 1070–71. Neale notes that the number of specific treatments for hydrophobia in the *Medical Digest* had risen from 29 to 45. Cf. J. S. Bristowe, 'Clinical Notes on Hydrophobia', *BMJ*, 1883, i: 760–61 and 808–10.
7. *BMJ*, 1886, i: 405–6.
8. *BMJ*, 1877, ii; 862; T. M. Dolan, *The Nature and Treatment of Rabies or Hydrophobia: Being the Report of the Special Commission Appointed by the Medical Press and Circular, with Valuable Additions*, London: Baillière, Tindall,

and Cox, 1878; T. M. Dolan, 'The Diagnosis of Hydrophobia', *Lancet*, 1880, i: 109–10; idem, 'The Prevention of Hydrophobia', *Lancet*, 1880, i: 184–85.

9. V. Mott, 'Rabies and How to Prevent It', *Journal of Social Science*, 1887, 22: 69.

10. This was a highly contentious claim, as none other than Jenner had allegedly stated that vaccination was of no value in the treatment of smallpox. *BMJ*, 1886, ii: 433–34.

11. C. Cameron, 'Micro-organisms and Disease', *BMJ*, 1881, ii: 583–87.

12. On the place of these individuals in debates over germ theories of disease see M. Worboys, *Spreading Germs: Disease Theories and Medical Practice in Britain, 1865–1900*, Cambridge: Cambridge University Press, 2000.

13. *Zoophilist*, 1881–1882, 1: 103. The correspondent went on to link antivivisection to antivaccination, worrying that as well as being vaccinated against smallpox, the public would have to suffer being porcinated against measles, equinated against glanders, caninated against rabies, and 'inoculated with sheep rot as a general specific for the gout'. 103.

14. 'Pasteur and Anthrax: Physiological Fallacies', *Zoophilist*, 1881–1882, 1: 239–40. This was said to have begun at the International Medical Congress in London in August 1881 and was continued, above all, by John Tyndall. See M. Worboys, *Spreading Germs*, 169–75.

15. The editorial argued that Koch had 'exposed the pretensions of vivisection'. *Zoophilist*, 1883–1884, 3: 25.

16. Ibid. Pasteur was one of the leading overseas speakers at the International Medical Congress in London in 1881. *Lancet*, 1881, ii: 297–98.

17. *Zoophilist*, 1883–1884, 3: 262 and 283.

18. *Zoophilist*, 1884–1885, 4: 99–100.

19. Ibid., 138.

20. *Medical Press*, 1884, i: 497–98 and 522–23; ii: 311–12; *Lancet*, 1884, ii: 77 and 335.

21. Geison, *Private Science*, 192.

22. A. T. Wilkinson, 'Prevention of Hydrophobia', *Medical Chronicle*, 1884–1885, 1: 47.

23. Geison, *Private Science*, 192.

24. *Zoophilist*, 1884–1885, 4: 99.

25. Ibid., 47.

26. Ibid.

27. Ibid.

28. Letters from *Lincoln Gazette*, 31 May 1884 and *Land and Water*, 28 June 1884, reprinted in *Zoophilist*, 1884–1885, 4: 53 and 83.

29. *Zoophilist*, 1884–1885, 4: 83, 99 and 105. *Animal World*, 16, 1885: 66, 75 and 81.

30. *Times*, 25 September 1885, 12b and 17 October 1885, 10c; *Veterinarian*, 1885, 58: 760–61.

31. *Lancet*, 1885, ii: 1054.

32. *Times*, 20 October 1885, 7d and 23 October 1885, 3d. Also see *Animal World*, 1886, 17: 13–14, 22, 82, 106 and 116.

33. *Times*, 28 October 1885, 5b and 9d. Jupille was referred to as 'Judith'. For an account of Pasteur's talk and its reception in France, see Geison, *Private Science*, 212–20; and P. Debré, *Louis Pasteur*, Baltimore: Johns Hopkins University Press, 1998, 435–46.

34. *Times*, 28 October 1885, 5b and 9d.

35. *The Graphic*, 7 November 1885, 502.

36. *Medical Press*, 1885, ii: 406, 453 and 478; *Lancet*, 1885, ii: 1054–55.
37. *Times*, 3 November 1885, 10c and 7 November 1885, 9f.
38. *Times*, 26 November 1885, 16f. Also see *Times*, 30 November 1885, 4c and *Veterinarian*, 1886, 59: 22–23, 87–88.
39. C. R. Drysdale, 'The Prevention of Rabies and Hydrophobia', *Daily News*, 21 December 1885, 6g.
40. Pasteur initially suggested that the attenuation of the virus was a quantitative, chemical change rather than qualitative, biological one; in other words, that the action of oxygen had reduced the concentration of the poison, not its virulence.
41. On Grancher's role see T. Gelfand, '11 January 1887, the Day Medicine Changed: Joseph Grancher's Defence of Pasteur's Treatment for Rabies', *Bulletin of the History of Medicine*, 2002, 76: 698–718.
42. *Illustrated London News*, 27 March 1886.
43. Ibid.
44. Hansen, 'America's First', 389–401.
45. The account of events was published across all columns, with illustrations, in the weekly *Bradford Illustrated Weekly Telegraph*, 20 March 1886, 1a–g.
46. *Bradford Telegraph*, 12 March 1886, 2f.
47. Subsequently, Hartley defended his remedy, claiming that his patient had not taken the remedy properly, nor had he followed ancillary recommendations such as avoiding alcohol. *Bradford Observer*, 23 March 1886, 7c.
48. *Daily News*, 15 March 1886, 5f; 17 March 1886, 5f; 19 March 1886, 5f.
49. *Bradford Observer*, 15 March 1886, 7d.
50. *Bradford Daily Telegraph*, 13 March 1886, 2f; 15 March 1886, 3c.
51. *London Daily News*, 18 March 1886, 5e; 20 March 1886, 5d.
52. *Bradford Daily Telegraph*, 17 March 1886, 2g.
53. *Bradford Daily Telegraph*, 18 March 1886, 2g.
54. *Bradford Daily Telegraph*, 15 March 1886, 3c.
55. *Bradford Daily Telegraph*, 20 March 1886, 2f and 3g.
56. *Bradford Daily Telegraph*, 15 March 1886, 3c.
57. *Animal World*, 1887, 18: 35.
58. Such visits to Continental Europe for the latest cures became more common in subsequent years, most notably with Robert Koch's Tuberculin remedy for tuberculosis in 1890. See C. Gradmann, 'Robert Koch and the Pressures of Scientific Research: Tuberculosis and Tuberculin', *Medical History*, 2001, 45: 1–32; Worboys, *Spreading Germs*, 224–28.
59. *London Daily News*, 17 March 1886, 5f.
60. F. E. Pirkis, 'Hydrophobia', *London Daily News*, 6 January 1886, 6f; *Bradford Observer*, 22 March 1886, 7d. Mitchell's treatment was paid for by F. E. Pirkis, though organised by Benjamin Bryan of the Victoria Street Society. Bryan worked closely with Frances Power Cobbe in the antivivisection movement and they co-authored a book on vivisection in the United States. F. Power Cobbe and B. Bryan, *Vivisection in America:. I. How it is taught; II. How it is practised*, London: Swan, Sonnenshein and Co., 1890.
61. J. H. Clarke, 'Pasteur and Hydrophobia', *Zoophilist*, 1885–1886, 5: 228.
62. *Times*, 10 July 1886, 13e.
63. *Bradford Daily Telegraph*, 8 July 1886, 5a.
64. *London Daily News*, 24 March 1886, 5e; 5 April 1886, 6c; 18 April 1886, 5g; 26 April 1886, 3a.

65. See Debré, *Louis Pasteur*, 444–45.
66. 'The Pasteur Craze', *Zoophilist*, 1885–1886, 5: 221–22. The editorial claimed that 'Pasteur's excuses and inconsistencies are of exactly the same kind as those of the vulgar quack'.
67. *London Daily News*, 13 April 1886, 3f.
68. *Report of Committee of Local Government Board to Inquire into Pasteur's Treatment of Hydrophobia*, Parl. Papers, 1887 [C.5087], lxvi, 429.
69. On the role of these doctors and scientists in the development of germ theories and practices in Britain, see Worboys, *Spreading Germs*, passim.
70. S. Paget, *Sir Victor Horsley: A Study of His Life and Work*, London: Constable and Co., 1919, 78–81; S. Paget, 'Horsley, Sir Victor Alexander Haden (1857–1916)', rev. Caroline Overy, *Oxford Dictionary of National Biography*, Oxford University Press, 2004 [http://www.oxforddnb.com/view/article/34000, accessed 18 Sept 2006].
71. T. M. Dolan, *Hydrophobia. M. Pasteur and His methods: A Critical Analysis*, London: H. K. Lewis, 1886; *Medical Press*, 1886, i: 515–16; *Lancet*, 1886, ii: 375. *Medical Chronicle*, 1886, 4: 433.
72. On Anna Kingsford see L. Williamson, 'Kingsford, Anna (1846–1888)', *Oxford Dictionary of National Biography*, Oxford University Press, 2004. [http://www.oxforddnb.com/view/article/15615, accessed 4 Dec 2006].
73. *Times*, 9 July 1886, 4e.
74. *Times*, 12 July 1886, 11f; *Medical Press*, 1886, ii: 36, 58, 76, 115, 135–36, 156–57, 204, and 340–41.
75. C. A. Gordon, *Inoculation for Rabies and Hydrophobia: A Study of the Literature of the Subject*, London: Baillière, Tindall and Cox, 1887; idem., *Comments on the Report of the Committee on M. Pasteur's Treatment of Rabies and Hydrophobia*, London: Baillière, Tindall and Cox, 1888; C. Pringle, 'Hydrophobia and the Treatment of the Bites of Rabid Animals by Suction', *Lancet*, 1886, i: 782–84; Ibid., 1225 and ii: 374–75; W. Curran, 'The Therapeutics of Hydrophobia', *Medical Press*, 1886, i: 284–86 and 308.
76. Quoted in *Medical Press*, 1886, ii: 115. V. Richards, *The Landmarks of Snake-Poison Literature: Being a Review of the More Important Researches into the Nature of Snake-Poisons*, Calcutta: Thacker, 1886, 20. This statement was also quoted in September 1886 by Thomas Dolan in his exchange with Victor Horsley. *BMJ*, 1886, ii: 476.
77. *Lancet*, 1886, i: 1225.
78. *Asclepiad*, 1889, 4: 290. On Richardson's view of bacteriology and immunology, see L. G. Stevenson, 'Science Down the Drain: On the Hostility of Certain Sanitarians to Animal Experimentation, Bacteriology and Immunology', *Bulletin of the History of Medicine*, 1955, 29: 1–26.
79. *Times*, 4 November 1886, 5d.
80. *Zoophilist*, 1889–1890, 9: 189.
81. *Times*, 25 October 1886, 7e.
82. The syndrome is now known as Guillain-Barré Syndrome (GBS) and is characterised by a spreading paralysis of the body. It is thought to be caused by an immune reaction to infection or vaccination.
83. Goffi's case was used by Pasteur's opponents in many countries, see *BMJ*, 1886, ii: 482–83.
84. Debré, *Louis Pasteur*, 458.

85. J. S. Bristowe and V. Horsley, 'A Case of Paralytic Rabies in Man', *Transactions of the Clinical Society*, 1888–1889, 22: 38–47.
86. *Sheffield Daily Telegraph*, 4 November 1886, 3d; 5 November 1886, 3f; 6 November 1886, 7c.
87. *Sheffield Daily Telegraph*, 11 November 1886, 3f.
88. Ibid. and *Lancet*, 1886, ii: 949. Thomas Hime was also interested, Wilde was potentially a case for observing the effects of passage of the virus through humans.
89. *Lancet*, 1887, i: 290.
90. *Zoophilist*, 1886–1887, 6: 161–62; C. T. Bell, 'Pasteur's Prophylactic', *Medical Press*, 1887, i: 491.
91. *BMJ*, 1886, ii: 475, 573, 602, 645, 753, 842, and 892.
92. Ibid., 573.
93. Ibid., 892.
94. *Medical Press*, 1886, i: 501. This point was also made in 1889 by Benjamin Ward Richardson. *Aesclepiad*, 1889, 4: 288–89.
95. *BMJ*, 1886, ii: 1158.
96. Gelfand, '11 January 1887', 698–718; *Medical Press*, 1887, i: 48, 82, and 128.
97. *Lancet*, 1887, i: 749–50, 1261; and ii: 440; *Medical Chronicle*, 1887, 11: 80–82; *Medical Press*, 1887, ii: 41–42.
98. This table was published regularly and became larger and larger, as deaths at all Pasteur Institutes across the world were counted. See *Zoophilist*, 1888–1889, 8: 189–90 and 1896–1897, 16: 158 et seq.
99. *Committee ... into Pasteur's Treatment*, 429.
100. Ibid., 434–35.
101. A. Lutaud, *Hydrophobia in Relation to M. Pasteur's Method and the Report of the English Committee*, London: Wittaker and Co., 1887. Also see J. H. Clark, *M. Pasteur and Hydrophobia: Dr Lutaud's New Work*, London: Victoria Street Society, 1887 and B. Bryan, *Reasons Why the English Committee's Report on Pasteurism does Not Settle the Question*, London: Victoria Street Society, 1887. Also see *Animal World*, 1887, 18: 130 and 178.
102. This estimate came from multiplying the average number of hydrophobia deaths in the ten years ending 1885 by 20; their assumption was that 1 in 20 of people bitten died. Ibid., 435.
103. *Lancet*, 1887, ii: 23–24. French critics read the Horsley Committee's report as 'confused and inconclusive'. Ibid., 235.
104. F. Karslake, *Rabies and Hydrophobia: Their Cause and Their Prevention*, London: W. & G. Foyle, 1919.
105. Ibid., 1–2.
106. M. A. Elston, 'Women and Anti-vivisection in Victorian England, 1870–1900', in N. A. Rupke, ed., *Vivisection in Historical Perspective*, London: Routledge, 1990, 259–94.
107. *Select Committee of House of Lords on Rabies in Dogs. Report, Proceedings, Minutes of Evidence, Appendix*, Parl. Papers, 1887, (322), xi, 451.
108. T. M. Dolan, *Hydrophobia: M. Pasteur and His Methods, a Critical Analysis*, London: H. K. Lewis, 1886; T. M. Dolan and C. B. Taylor, 'Is Pasteurism a Fraud?', *Contemporary Review*, 1890, 2: 29; idem, *Pasteur and Rabies*, London: George Bell and Sons, 1890.
109. *BMJ*, 1888, ii: 460–61.

110. *The Globe*, 29 June 1889.
111. V. Horsley, *On Hydrophobia and Its Treatment, Especially by the Hot-Air Bath, Commonly Termed the Bouisson Remedy*, London: J. Bale & Sons, 1888; Paget, *Sir Victor Horsley*, 82–84.
112. *Lancet*, 1889, i: 248 and 402.
113. These included the pathologist J. G. Adami, see *BMJ*, 1889, ii: 808–10.
114. *Times*, 17 June 1889, 5e.
115. Wellcome Library Archives, SA/LIS/C2, Mansion House Fund Meeting.
116. *Lancet*, 1889, ii: 22.
117. H. Chick, *War on Disease: A History of the Lister Institute*, London: André Deutsch 1971, 23–27. Pasteur had corresponded with Armand Ruffer in the summer of 1889 and suggested Ruffer to organise a London antirabies institute. Wellcome Library Archives, SA/LIS/C3–5. Also see speeches given at the Lord Mayor's Meeting. SA/LIS/C2. Also see A. Ruffer, 'Rabies and Its Preventive Treatment', *Journal of the Royal Society of Arts*, 1889, 38: 30–39.
118. *Times*, 8 January 1890, 8b.
119. *Medical Press*, 1889, ii: 41.
120. We have assumed 1 in 15 rather than the 1 in 20 used by Horsley's Committee. Contemporary estimates varied between 1 in 10 and 1 in 20.
121. See Graph 4.1.
122. *Stalybridge Reporter*, 1 February 1890, 5f. The incident was also reported in the *Denton and Hyde Examiner*, 1 February 1890, 5c. The case was later reported in full in the *BMJ*, 1890, ii: 149.
123. *North Cheshire Herald*, 1 February 1890, 5d.
124. *Stalybridge Reporter*, 1 February 1890, 5f.
125. Dr George Kisby of the Manchester Royal Infirmary (MRI) was recruited for the task and left for Paris in the early afternoon with the dog's brain and spine. An account of Pasteur's work had been given to the Lancashire Veterinary Medical Association on 8 December 1886; see J. B. Wolstenholm, 'Rabies and hydrophobia', *Veterinarian*, 1887, 60: 57–63.
126. Frederick H. Westmacott had just qualified in 1890. After developing St John's Hospital for the Ear in Manchester, he went on to secure an Honorary appointment at the MRI and to be a lecturer at the University of Manchester. 'Frederick H. Westmacott', E. B. Brockbank, ed., *Honorary Medical Staff of the Manchester Royal Infirmary, 1830–1948*, Manchester: Manchester University Press, 1965, 174–76.
127. Tameside Record Office, CA Sta/10. *Medical Officer's report of the births and deaths for the 53 weeks ending in 3 January 1891*. Chamberlayne reported that 115 dogs were seized, of which 42 were returned to owners, 49 were given away or sold and 24 were destroyed, and the owners of 22 dogs were summoned, of whom 16 were fined.
128. John Rushton Hartley was 43 years of age in 1890. In the 1880s he ran the co-operative shop in Colne. His remedy was sold as 'Hydrophobine'. *Colne and Nelson Times*, 1 June 1889, 4c. We would like to thank Margaret Heap for valuable information on Hartley.
129. 'Obituary: George W. Sidebotham', *BMJ*, 1910, i: 1090. Two further mad dog victims from Hyde were taken to Paris for treatment in 1893 – one of whom died nine months later.

130. H. Valier, *The Politics of Scientific Medicine in Manchester, c.1900–1960*, Unpublished PhD, University of Manchester, 2002, 67–86.
131. Ashton was also a major backer of Owens College, soon to be the Victoria University of Manchester and was himself the graduate of a German University. Jane Bedford, 'Ashton, Thomas (1818–1898)', *Oxford Dictionary of National Biography*, Oxford University Press, 2004 [http://www.oxforddnb.com/view/article/50518, accessed 4 Dec 2006]
132. It is a moot point whether arranging for mad dog victims to be sent to Pasteur was a vote winner. A number of anti-vivisectionists celebrated in 1895 when Roscoe lost his parliamentary seat. *The Zoophilist*, 1895, 15: 219.
133. *Stalybridge Reporter*, 8 February 1890, 6f.
134. F. Westmacott, *Manchester Courier*, August 1890, cutting in John Rylands University Library Manchester, Manchester Medical Collection, MMC/1/WestmacottF. The local papers reported the courage of the patients, with Pasteur himself allegedly giving William Schofield a ten centime piece and telling him that the way he had borne his injections showed that 'He is a Briton'.
135. *Cheshire Herald*, 1 March 1890, 6f.
136. Tameside Archives: CA Sta. 297/10. *Medical Officer's report of the births and deaths in the 53 weeks ended 3 January 1891* (1891), 33. This report contained no mention of the four people bitten or their treatment.
137. *Clitheroe Times*, 24 May 1890, 3f.
138. *Heywood Advertiser*, 13 June 1890, 8a. Hartley's treatment was also reported in the anti-vivisection press. *Zoophilist*, 1889–1890, 9: 142–43.
139. Hartley had a treatment book with records of over 500 cases with no failures at all. Known locally as the 'Mad Dog Man'. His secret remedy was mentioned in the *Gentleman's Magazine* in 1753 and had been in his family since, being passed on to him by his grandfather. *Gentlemen's Magazine*, 1753, 23: 368.
140. *Stalybridge Reporter*, 28 June 1890, 5e.
141. Ibid.
142. 'John Edward Platt (1865–1910)', Brockbank, *Honorary Medical Staff*, 123–25.
143. *Heywood Advertiser*, 4 July 1890, 6a.
144. *North Cheshire Herald*, 28 June 1890, 5f.
145. *Lancet*, 1890, i: 980 and 1436.
146. *North Cheshire Herald*, 7 September 1895, 3c, 4 d–e, and 8a; Also 14 September 1895, 4c–d.
147. This statement was reported in the *North Cheshire Herald*, 7 September 1895, 4d, but was not included in *Hansard*; *Parliamentary Debates*, 1895, 136: 435–36.
148. M. Mulvihill, 'Mad Dogs and Scotsmen: A Plain Tale from the Military Collections of the India Office Records Section of the British Library', *eBJL*, 2006, Article 6. India Office Records, IOR/MIL/7/7379, Military Collections. We would like to thank Dr Pratik Chakrabarti, University of Kent, for this information.
149. *Lancet*, 1901, i: 759.
150. IOR, Home Department Proceedings, Branch – Medical, Vol. 2, July–December 1900, P/1203, Letter from Lord Curzon and Council to Secretary of State for

India, 7 June 1900. We would like to thank Dr Pratik Chakrabarti for this information.
151. *Lancet*, 1896, i: 261.
152. *BMJ*, 1890, i: 149.
153. The *Zoophilist* listed the following deaths in the 1890s. In 1894: Enoch Oakes (Rochdale), Malcolm Stephenson (Birkenhead), James Bentley (Clayton-le-Moors), Benjamin Howard (Hyde), Samuel Nothard (Todmorden), Ethel Wilkins (Twickenham), Thomas Lambert (Stockport), and Thomas Openshaw (Radcliffe); In 1896: James Thompson (Reddish). Two brothers from Ireland died in 1897: Terence and John McQuilliam.
154. *Lancet*, 1896, i: 261.
155. F. J. Allen, 'Personal Experience of the Pasteur Anti-Rabic Treatment', *Birmingham Medical Review*, 1898, 43: 1–15; E. H. Julian, 'Saved from Hydrophobia', *World Wide Magazine*, April 1899.
156. O. Beatty Kingston, 'A Child's Experiences in M. Pasteur's Institute', *Woman at Home*, 1894, 1: 39.
157. Ibid., 40.
158. *Lancet*, 1895, ii: 886–90 and 933–36.

5 Rabies Banished: Muzzling and Its Discontents, 1885–1902

1. In a letter read by Henry Roscoe at the Mansion House meeting on 1 July 1889, Pasteur's statement highlighted how wolves lurked around the question of rabies; he wrote, 'Let England which has exterminated its wolves, make a vigorous effort and it will succeed in extirpating rabies'. *Lancet*, 1889, ii: 22.
2. S. Paget, *Sir Victor Horsley: A Study of His Life and Work*, London: Constable and Co., 1919, 84–89.
3. *Dog Owners' Annual for 1891*, London: Dean and Son, 1891, 174.
4. Ibid.
5. Many contemporary veterinarians and doctors argued that both diseases were over reported. See J. Rose Bradford, 'Two Lectures on Rabies', *Lancet*, 1900, i: 593–96 and 578–81.
6. On the crisis see A. J. Sewell, 'Rabies', *Kennel Gazette*, April 1886, 75–76.
7. *Times*, 17 February 1885, 7f; 20 February 1885, 6f; 25 September 1885, 12b; *Lancet*, 1885, ii: 812–13 and 828.
8. 'Deaths from Hydrophobia', *Times*, 30 October 1885, 6c; *Daily News*, 23 November 1885, 6e.
9. *Times*, 3 November 1885, 10c.
10. See letters: *Times*, 19 October 1885, 6f; 30 November 1885, 4c; 26 November 1885, 16f; also see C. R. Drysdale, *Daily News*, 21 December 1885, 6g.
11. *Commissioner of Police of Metropolis, Report, 1885*, Parl. Papers, 1886 [C.4823], xxxiv, 329–80. On the problems posed by the disposal of thousands of dog carcases see *Daily News*, 1 February 1886, 2d and 26 February 1886, 7b. Battersea Dogs Home received 1,438 dogs in October, which was typical, but the number rose to 3,775 in November, 6,536 in December, and 4,305 in January 1886. Numbers did not return to normal until November 1886.

12. *Select Committee of House of Lords on Rabies in Dogs. Report, Proceedings, Minutes of Evidence, Appendix*, Parl Papers, 1887, (322), xi, 527.
13. *Animal World*, 1886, 17: 13–14.
14. *Animal World*, 1886, 17: 117.
15. H. Ritvo, *The Animal Estate: The English and Other Creatures in the Victorian Age*, Cambridge, MA: Harvard University Press, 1987.
16. *Veterinarian*, 1886, 59: 22–23. Frances Power Cobbe, the leading antivivisectionist, had long complained about the RSPCA's acceptance of the Vivisection Act, 1876, now she complained about its failure to oppose muzzling. *Times*, 18 October 1886, 4a.
17. *Times*, 12 July 1886, 10e.
18. *Times*, 14 July 1886, 8a.
19. *Times*, 9 June 1886, 10e; 14 June 1886, 3d; 30 June 1886, 4b; 12 July 1886, 3f and 10e; 14 July 1886, 4b; 30 July 1888, 4f; 13 August 1886, 6e.
20. *Dog Owners' Annual for 1891*, London: Dean and Son, 1891, 174–78; *Times*, 1 November 1886, 13a; 23 December 1886, 7d.
21. Times, 6 September 1886, 14a.
22. *Animal World*, 1886, 17: 116 and 143.
23. *Kennel Gazette*, November 1886, 249.
24. *Select Committee … on Rabies in Dogs*, 557–58.
25. Both orders were reprinted in *Minutes of Evidence Taken Before the Dept. Committee to Inquire into Working of Laws Relating to Dogs*, Parl. Papers, 1897 [C.8378], xxxiv, 173–78.
26. J. Ablett, *The Muzzling Fiasco*, London: S. E. Stanesby, 1897.
27. Ibid., 4.
28. *Select Committee … on Rabies in Dogs*, 451 et seq. There were accusations that the Committee was 'stuffed' with country gentlemen and that it was bound to recommend against universal muzzling in order to protect hunting with hounds.
29. C. A. Gordon, 'The Prevention of Rabies and Hydrophobia', *Lancet*, 1890, 1: 1148–49.
30. *Select Committee … on Rabies in Dogs*, 630–31.
31. A good example are the literary trio of Gordon Stables, Frederick Pirkis, and Catherine Pirkis. G. S. Woods, 'Stables, William Gordon (1840–1910)', rev. Guy Arnold, *Oxford Dictionary of National Biography*, Oxford University Press, 2004 [http://www.oxforddnb.com/view/article/36229, accessed 3 Jan 2007]. On Pirkis's see the brief biographical note in D. G. Greene, ed., *Detection by Gaslight: 14 Victorian Detective Stories*, London: Dover, 1997, 45–46.
32. J. Woodruffe Hill, *The Management and Diseases of the Dog*, London: Baillière, Tindall, and Cox, 1900, 240.
33. *Select Committee … on Rabies in Dogs*, 457.
34. Ibid., 454.
35. Ibid., 455.
36. *Kennel Gazette*, September 1889, 219.
37. Ibid.
38. *Select Committee … on Rabies in Dogs*, 455. Details of measures in France, Switzerland, Austria, and New York were published in the Appendix, 708–20.
39. Ibid., 650–55.
40. *Kennel Gazette*, July 1889, 167.

41. *Times*, 9 October 1889, 3a–c.
42. *Times*, 8 October 1889, 6f.
43. *Times*, 9 October 1889, 3a.
44. *Times*, 1 August 1889, 7d.
45. *Times*, 28 October 1889, 4b. On Horsley and Cobbe see Paget, *Sir Victor Horsley*, 147.
46. *Times*, 22 October 1889, 13b.
47. *Animal World*, 1890, 21: 31.
48. *Times*, 4 January 1890, 7f.
49. *Dog Owner's Annual*, 1891, 180–81.
50. R. J. Olney, 'Chaplin, Henry, first Viscount Chaplin (1840–1923)', *Oxford Dictionary of National Biography*, Oxford University Press, 2004 [http://www.oxforddnb.com/view/article/32363, accessed 3 Jan 2007].
51. *Times*, 16 December 1889, 7c.
52. *Animal World*, 1890, 21: 35.
53. *Animal World*, 1890, 21: 2.
54. Ibid.
55. Ibid., 31.
56. *Times*, 15 January 1890, 13d and 23 January 1890, 8b.
57. This outbreak had a silver lining for the pro-muzzlers. The Medical Officer of Health for Nottingham was Arthur Whitelegge, Victor Horsley's brother-in-law, and it was with samples of rabid dog brain from Nottingham that Horsley confirmed Pasteur's work for the 1887 Departmental Committee. Steve Sturdy, 'Whitelegge, Sir (Benjamin) Arthur (1852–1933)', *Oxford Dictionary of National Biography*, Oxford University Press, 2004 [http://www.oxforddnb.com/view/article/38123, accessed 3 Jan 2007].
58. One supporter of Chaplin wrote to the Standard that he was 'better qualified to judge than ninety-nine out of a hundred of his countrymen. The "Squire of Blakeney" is no red-tape potentate – he know, and has known, dogs and their ways ever since he was able to walk, and whatever the feeling of certain malcontents in Kent, it is certain that in Lincolnshire the owner of Hermit is recognised as well-nigh infallible in connection with all departments of the world of sport.' Quoted in *Animal World*, 1890, 21: 36.
59. DOPA, *The Kent Dog Owners and the New Muzzling Order*, London: DOPA, 1890. *Animal World*, 1890, 21: 36.
60. *Standard*, 6 January 1890, 2e–f; 9 January 1890, 2c.
61. Ibid., 4 January 1890, 2d; 7 January 1890, 2c–d. Also see DOPA, *The Kent Dog Owners*, 29 and 36; *Stock Keeper and Furriers' Chronicle*, 10 January 1890.
62. *Manchester Guardian*, 6 January 1890, 3f; 8 January 1890, 2d; 9 January 1890, 8d–e; 14 January 1890, 9f–g; 18 January 1890, 5a and 9h–10a; 25 January 1890, 9f–g.
63. Ibid., 9 February 1890, 9f.
64. *Sheffield Daily Telegraph*, 3 January 1890, 6d; 6 January 1890, 8a; 8 January 1890, 6f.
65. Ibid., 7 January 1890, 7d and 9 January 1890, 8b.
66. Ibid., 36.
67. *Veterinarian*, 1889, 62: 550.

68. G. Fleming, 'The Suppression of Rabies in the United Kingdom', *Nineteenth Century*, 1890, 26: 497–512.
69. *Annual Report of Veterinary Department for 1890*, Parl. Papers, 1890–1891 [C.6533], xxv, 92–94.
70. *Times*, 14 January 1891, 9d and 13e. This scheme was termed the 'scientific plan'.
71. *Annual Report of Veterinary Department*, 1890, 93–94. In fact, 206 out of 229 cases were in these three counties. 'Beth Gelert' wrote again to the press on behalf of dogs. *Lancet*, 1894, i: 556.
72. *Times*, 9 July 1897, 10b. The trend had been upwards since 1893, with the actual figures of prosecutions and convictions (in brackets) for each year as follows: 1893: 3,539 (2,952); 1894: 9,265 (7,803); 1895: 31,434 (27,522); 1896: 25,046 (21,390).
73. *Animal World*, 1897, 28: 78.
74. *Annual Reports of Proceedings Under Contagious Diseases (Animals) Acts, and Markets and Fairs (Weighing of Cattle) Acts, 1896*, Parl. Papers, 1897 [C.8389], xxiii, 49–50.
75. Ibid., 62.
76. *Times*, 29 January 1896, 9e and 7 February 1896, 11b; *Animal World*, 1896, 27: 34–35 and 50.
77. *Kennel Gazette*, December 1896, 358. *Times*, 4 February 1896, 11c–e and 5 February 1896, 11d.
78. *Ladies Kennel Journal*, 1896, 3: 174.
79. *Our Dogs*, 3 April 1897, 461; *Kennel Gazette*, December 1897, 460.
80. *Departmental Committee to Inquire into Working of Laws Relating to Dogs Report*, Parl. Papers, 1897 [C.8320], xxxiv, 1–16, *Minutes of Evidence Taken Before the Dept. Committee*, 17–231.
81. On Long's time at the Department of Agriculture see W. Long, *Memories: By the Right Honourable Viscount Long of Wraxall, F. R. S. (Walter Long)*, London: Hutchinson, 1923, 114–33.
82. Alvin Jackson, 'Long, Walter Hume, First Viscount Long (1854–1924)', *Oxford Dictionary of National Biography*, Oxford University Press, 2004 [http://www.oxforddnb.com/view/article/34591, accessed 3 Jan 2007].
83. Long, *Memories*, 118.
84. *Times*, 1 Feb 1897, 6f and 3 February 4e.
85. Quoted in *Kennel Gazette*, February 1897, 49.
86. *Standard*, 7 April 1897, 3a–b.
87. E. Millais, *Times*, 20 April 1897, 10b.
88. Long, *Memories*, 121–22.
89. *Kennel Gazette*, 1897, 18: 181.
90. *Our Dogs*, 24 April 1897, 494.
91. *Standard*, 10 April 1897, 2b.
92. *Our Dogs*, 1 May 1897, 521 and 29 May 1897, 645; *Animal World*, 1897, 28: 66, 80 and 131–32. The Society argued that identification with a collar and tag would be more effective than muzzling.
93. *Standard*, 14 April 1897, 2d.
94. Also see Ibid., 15 April 1897, 2f.
95. 'Editorial', *Our Dogs*, 29 May 1897, 645.
96. *Our Dogs*, 26 June 1896, 752 and 17 July 1896, 836.

97. *Hansard*, 53, 1898, 903 and 54, 1898, 280–1, 488 and 834.
98. Long, *Memories*, 128–29.
99. *Times*, 8 December 1898, 7e.
100. *Our Dogs*, 28 August 1897, 102–3.
101. *Our Dogs*, 1 May 1897, 521–23.
102. *Our Dogs*, 29 May 1897, 645–47.
103. 'Muzzles for Ladies', *Standard Magazine*, 1897, 20: 485–89.
104. *Times*, 15 April 1997, 12d; *Kennel Gazette*, April 1897, 143. There were only eight cases of rabies in London the first six months of 1897. *Times*, 30 June 1897, 13a.
105. *Times*, 27 October 1896, 9f and 14 April 1897, 12d.
106. *Times*, 23 April 1897, 4e.
107. *Times*, 17 August 1898, 5f. Cf. *Our Dogs*, 19 August 1899, 168.
108. *Our Dogs*, 28 August 1899, 102–3; *Times*, 31 August 1899, 10d and 28 August 1901, 12f. In 1901 Pirkis noted the view of the Manchester veterinarian W. G. R. A. Cox that only around 5 per cent of reports were confirmed by laboratory tests.
109. *Our Dogs*, 17 July 1897, 836.
110. *Animals*, August 1899.
111. *Zoophilist*, 1897–1898, 17: 1. In his autobiography Long stated that 'I have often received great commendation for the policy which was carried out during those five to six years. Really it was undeserved. I had nothing to do except sit tight, tell Major Tennant and his officials to carry out the regulations, and defend the policy in Parliament and in the country.' Long, *Memories*, 120.
112. *Times*, 19 November 1897, 6e.
113. *Times*, 24 April 1897, 11c.
114. *Ladies Kennel Journal*, September 1897, 103.
115. *Our Dogs*, 24 May 1897, 494.
116. *Times*, 21 August 1901, 7c.
117. *Times*, 19 November 1897, 9b, also see G. Sims Woodhead, 'Royal Veterinary College: Introductory Address', *Lancet*, 1899, ii: 957.
118. *Times*, 19 November 1897, 6e.
119. *Times*, 17 July 1897, 10c.
120. *Annual Reports of Proceedings under Contagious Diseases (Animals) Acts, and Markets and Fairs (Weighing of Cattle) Acts, 1897*, 1898 [C.8796] xx: 81–2 and 84–109.
121. *Times*, 29 January 1896, 9e. There was a large trade, especially of red setters into Liverpool and Fleetwood.
122. *Report of Proceedings Under the Diseases of Animals Act, 1894, as Regards Ireland, for 1905*, Parl. Papers, 1906, [Cd. 3133], cxi, 293.
123. Evidence of Henry Fitzgibbon, MD, Member of Lord de Vesci's Committee, *Minutes of Evidence taken before the Dept. Committee*, 133.
124. *Times*, 20 April 1897, 10b.
125. *Kennel Gazette*, August 1899, 334.
126. *Times*, 31 August 1899, 10d.
127. *Hansard*, 75, 1899, 1290; *Times*, 4 August 1899, 4c.
128. *Our Dogs*, 19 August 1899, 167. An editorial noted that they had not liked his methods, but went on, 'we shall not withhold credit', acknowledging Long's bravery when subjected to 'ridicule and abuse'. *Lancet*, 1899, ii: 956.

129. I. Griffiths, 'The Pontudulais Mad Dog', [http://genuki.org.uk/big/wal/ CMN/Llanedi/Education.html, accessed 30 March 2005]. Eight children were bitten and sent to the Pasteur Institute. Cf. *Zoophilist*, 1898–1899, 18: 123.

130. *Annual Reports of Proceedings Under Contagious Diseases (Animals) Acts, and Markets and Fairs (Weighing of Cattle) Acts, 1899*, Parl. Papers, 1900, [Cd.107], xvii, 61.

131. *Annual Reports of Proceedings Under the Diseases of Animals Acts, the Markets and Fairs (Weighing of Cattle) Acts, &c., for 1900*, Parl. Papers, 1901, [Cd. 535], xvii, 85–6 and 105–6.

132. *Times*, 21 August 1901, 7c.

133. *Annual Reports of Proceedings Under the Diseases of Animals Acts, the Markets and Fairs (Weighing of Cattle) Acts, &c., for 1901*, Parl. Papers, 1902 [Cd. 1041], xx, 33; *Annual Reports of Proceedings Under the Diseases of Animals Acts, the Markets and Fairs (Weighing of Cattle) Acts, &c., for 1902*, Parl. Papers, 1903, [Cd. 1520], xvii, 157.

134. *The Scotsman*, 25 November 2002.

135. *Diseases of Animals Acts, the 1901*, 18.

136. E. C. Miller, 'Trouble with She-Dicks: Private Eyes and Public Women in the Adventures of Loveday Brooke, Lady Detective', *Victorian Literature and Culture*, 2005, 33: 47–65.

137. Ibid., 63.

138. N. Clausson, 'Degeneration, Fin-de-Siecle Gothic, and the Science of Detection: Arthur Conan Doyle's *The Hound of the Baskervilles* and the Emergence of the Modern Detective Story', *Journal of Narrative Theory*, 2005, 35: 60–87.

139. A. Conan Doyle, *Hound of the Baskervilles*, Harmondsworth: Penguin Classics, 2003, 18.

140. Mortimer insisted that Sir Charles died of terror and fright, affected by the alleged curse that haunted the Baskerville family through the generations. Sir Charles Baskervilles was, as Mortimer described, a man obsessed with the idea of a dog continually haunting him, taunting his mind, and harassing him from afar. So overbearing this incessant infatuation became that it ultimately led to his untimely death by straining his nervous system, 'to breaking point', because 'he had taken the legend exceedingly to heart, so much so that, although he would walk in his own grounds, nothing would induce him to go out upon the moor at night'.

141. Doyle, *Hound*, 130

142. Ibid.

143. Ibid., 148.

144. Ibid., 150.

145. *Our Dogs*, 26 August 1899, 237.

146. *Times*, 31 August 1899, 10d.

147. Sir H. Smith, 'More About Retrievers', *Blackwood's Magazine*, 1900, 168: 212–14. Smith repeated these points in his autobiography. H. Smith, *From Constable to Commissioner*, London: Chatto and Windus, 1910, 77–79.

148. *Annual Reports of Proceedings Under Contagious Diseases (Animals) Acts, and Markets and Fairs (Weighing of Cattle) Acts, 1898*, Parl. Papers, 1899 [C.9208], xviii, 56.

6   Rabies Excluded: Quarantines to Pet Passports, 1902–2000

1. *Times*, 3 December 1903, 4f. W. Long, *Memories: By the Right Honourable Viscount Long of Wraxall, F.R.S. (Walter Long)*, London: Hutchinson, 1923, 130.
2. *The New Age*, 691, 1907: 101.
3. *Annual Report of Proceedings Under the Disease of Animals, Markets and Fairs Acts for 1903*, Parl. Papers, 1904 [Cd.2006], xvi, 61.
4. *Report of the Commissioner of Police for the Metropolis for 1905*, Parl. Papers, 1906 [Cd.3180], xlix, 386.
5. *Daily Mirror*, 17 September 1906, 6c; *Report of the Commissioner of Police for the Metropolis for 1914*, Parl. Papers, 1916 [Cd.8188], xlv, 566.
6. *Times*, 23 April 1904, 11b; *Daily Mirror*, 22 March 1906, 5c.
7. Ibid., 26 December 1908, 9c.
8. *Daily Mirror*, 5 March 1908, 4b.
9. *Times*, 26 March 1910, 10e and 28 March 1910, 4e.
10. Zuckerman Archive, University of East Anglia. SZ/CIR/3, CIR/1 Appendix IV.
11. SZ/CIR/3 CIR/1, Appendix IV. Letter to Sir Walter Long from H. C. W. Varney, Chief Veterinary Office, 24 October 1914.
12. Dog shows were allowed from February 1918. *Times*, 9 February 1918, 7b.
13. E. Le Chene, *Silent Heroes: An Animals' Roll of Honour*, London: Souvenir Press, 1997.
14. *Lancet*, 1919, i: 806.
15. *Hansard*, HC, 1918, 110: 604–6, 2878–79 and 3211.
16. *Western Evening Herald*, 22 November 1918, 3e–f. The Committee was comprised of: Lord Clinton (Chair), George Lambert, MP, and Major-General Sir David Bruce. *Times*, 6 January 1919, 3e. Retrospective assessments suggested that the disease may have been present as early as May 1918. *Western Evening Herald*, 22 November 1918, 3e–f.
17. Hansard, HC, 105: 1971–72 and 2138; 106: 44.
18. *Western Evening Herald*, 16 September 1918, 1h.
19. Ibid., 25 September 1918, 3d.
20. *Illustrated Western Weekly News*, 12 October 1918 and 16 November 1918, 2e–f, 9h; *Western Evening Herald*, 10 October 1918, 1d; *Western Daily Mercury*, 16 October 1918, 6c; *Express and Echo*, 28 November 1918, 2e; 7 April 1919, 1g.
21. *Hansard*, HC, 1919, 112: 2012–18.
22. *British Medical Journal (BMJ)*, 1919, ii: 418.
23. *Western Daily Mail*, 19 January 1919, 1f; *Express and* Echo, 24 April 1919, 3g.
24. *Western Evening Herald*, 26 October 1918, 3e; *Western Evening Herald*, 4 January 1919, 3f and 14 January 1919, 4b. In total 21 people from Devon and Cornwall were treated in Paris, though a number who went were refused treatment because there was no evidence of a dog bite. *Annual Report of the Medical Officer of the Local Government Board, Report for 1918–19*, Parl. Papers, 1919 [Cd.462], xxiv, 684.
25. *Lancet*, 1919, i: 259 and 350–51. The last case in Cornwall was on 11 February 1919 and first cases in Newport and Cardiff on 13–14 March 1919.
26. In 1909 the Viceroy, Lord Minto, was bitten by a stray dog and underwent precautionary treatment at Kasauli. *Daily Mirror*, 15 May 1909, 5c. Semple had developed a killed vaccine for use in India and it was this that he made

available in London from 1 June 1919. *Annual Report ... for 1918–19* [Cmd.462], 683. *Annual Report of the Medical Officer of the Local Government Board, Report for 1919–20*, Parl. Papers, 1920 [Cmd.978], xvii, 920–33.

27. *Daily Express*, 17 April 1919, 1d; *Daily Mirror*, 17 April 1919, 2a–b; *Times*, 17 April 1919, 12f and 19 April 1919, 7a.
28. *Daily Mail*, 21 April 1919, 1a; *Daily Express*, 19 April 1919, 1a–b.
29. *Daily Express*, 19 April 1919, 1a–b; *Daily Mirror*, 19 April 1919, 2a and 8–9.
30. *Daily Herald*, 23 April 1919, 2a; For photographs of the Dogs' Home see *Daily Express*, 24 April 1919, 3a–h.
31. *Daily Mirror*, 21 April 1919, 2a.
32. *Daily Express*, 19 April 1919, 1a–b; 22 April 1919.
33. *Times*, 21 April 1919, 8f.
34. *Daily Herald*, 25 April 1919, 3h; *Daily Express*, 24 April 1919, 5c.
35. *Daily Express*, 24 April 1919, 5c–e.
36. *Daily Mail*, 24 April 1919, 2d. Orders were lifted in most parts of London in January 1921.
37. *Times*, 10 May 1919, 9c; 13 May 1919, 7e. The Kennel Club favoured universal muzzling because it would then be unnecessary to restrict the movement of dogs, hence, dog shows could start again.
38. F. Karslake, *Rabies and Hydrophobia: Their Cause and Their Prevention*, London: W & G Foyle, 1919; E. Douglas Hume, *Hydrophobia and the Mad Dog Scare: An Exposure of Pasteur and Pasteurism, etc.*, London, 1919.
39. *Lancet*, 1921, i: 727.
40. At the same time there was another animal cruelty Bill going through Parliament – the Anaesthetics for Animals Act. *Hansard*, HC, 1919, 113: 2435–80; 114: 1627–52; 116: 725–868; 117: 511–34. This required veterinarians and others operating on animals to use anaesthetics, the irony being that anaesthesia was compulsory for most vivisection experiments, but was not required, nor seeming widely used, in veterinary operations, including common practices such as tail docking and castration. *Times*, 3 May 1919, 16a. Also see T. Tansey, 'Protection Against Dog Distemper and Dogs protection Bills; The Medical Research Council and Anti-vivisectionist protest, 1911–1933', *Medical History*, 1994, 38: 1–26.
41. *Royal Commission on Vivisection, Final Report*, Parl. Papers, 1912–1913, xlviii [Cd.6114], 401–540. Victor Horsley continued to be a hate figure amongst antivivisectionists. See *Anti-Vivisection Review*, 1910–1911, 2: 139 and illustration.
42. *Times*, 11 April 1919, 8b. The fate of Lord Doneraile, the Irish aristocrat who had died after treatment at the Pasteur Institute in 1886, was again debated. Antivivisectionists argued that his groom, who had also been bitten, survived because he did not go to Paris, however, supporters of the treatment pointed out that the groom had gone with his master. *Times*, 28 April 1919, 8c and 30 April 8a.
43. *Times*, 17 April 1919, 8b.
44. The fact that dogs were no longer used in rabies work allowed antivivisectionists to ask if it had ever been necessary, and why vivisectors objected to restrictions which would have no impact on their work. *Times*, 3 May 1919, 8c.
45. *Times*, 23 April 1919, 13c.
46. *Church Times*, 11 July 1919. Reprinted in the *Animal Defender*, August 1919.

47. *Lancet* 1919, i: 958. *Hansard*, HC, 1919, 116: 725–868.
48. In this discussion rabies was not mentioned, the examples of the value of research on dogs were the wider national interest – gas warfare and rickets in children. *Lancet*, 1919, ii: 39–40. *Hansard*, HC, 1919, 117: 511–14.
49. *Southern Daily Echo*, 8 January 1921, 3d; 15 January 1921, 3f; 26 January 1921, 3e,
50. *Southern Daily Echo*, 4 May 1922, 3d; *Veterinary Record*, 1922, 34: 353, 385, 408.
51. *Times*, 22 February 1921, 7d.
52. *Times*, 4 September 1923, 7b.
53. Sir F. Hobday, *Fifty Years a Veterinary Surgeon*, London: Hutchinson & Co, 1938, 35. We thank Dr Abigail Woods for this reference.
54. Ibid., 34.
55. *Times*, 27 July 1926, 17d.
56. *Veterinary Record*, 5 May 1927.
57. *Hansard*, HC, 1927, 203: 835.
58. *Times*, 25 February 1927, 15f.
59. *British Medical Journal (BMJ)*, 1924, i: 1049–50.
60. A. Woods, *A Manufactured Plague: Foot and Mouth Disease in Britain*, London: Earthscan, 2004.
61. League of Nations, International Rabies Conference, Genève, 1927.
62. Ibid., 57. Also see *Times*, 6 July 1926, 15b and 4 May 1927, 15d. *Hansard*, HC, 208, 1927: 571–72.
63. A. A. King, A. R. Fooks, M. Aubert and A. I. Wandeler, eds, *Historical Perspective of Rabies in Europe and the Mediterranean Basin*, Paris: l'Organisation mondiale de la santé animale (OIE), 2004, 79–98 and 129–46.
64. 'Kasauli of the Hills an Indian Pasteur Institute., Mad Beasts And Men', *Times*, 25 April 1919, 11a.
65. D. Semple, *The Preparation of a Safe and Efficient Antirabic Vaccine*, Calcutta: Supt. Govt. Print. India, 1911.
66. On the work of the Pasteur Institute for Southern India in 1923–1924, see *BMJ*, 1924, ii: 1135.
67. S. C. J. Bennett, 'A Note on the Nature of the Rabies Virus in the Anglo-Egyptian Sudan', *Veterinary Journal*, 1927, 83: 123–26; R. A. B. Stanhope, 'Rabies in Malaya', *Veterinary Record*, 1928, 8: 997–1004 and 1005.
68. E. Weston and J. L. Pawan, 'An Outbreak of Rabies in Trinidad Without History of Bites, and with Symptoms of Acute Ascending Myelitis', *Lancet*, 1931, ii: 622–28.
69. *Times*, 17 November 1931, 14f; *New York Times*, 29 September 1931, 8f; also see 5 September 1931, 13b. For a review of the episode see E. De Verteuil and F. W. Urich, 'The Study and Control of Paralytic Rabies Transmitted by Bats in Trinidad, British West Indies', *Transactions of the Royal Society of Tropical Medicine*, 1935, 29: 317–47.
70. L. Wilkinson, 'The Development of the Virus Concept as Reflected in Corpora of Studies on Individual Pathogens: 4. Rabies – Two Millennia of Ideas and Conjecture on the Aetiology of a Virus Disease', *Medical History*, 1977, 21: 15–31; idem, 'History', in A. C. Jackson and W. H. Wunner, eds, *Rabies*, Amsterdam: Rodopi, 2002, 1–22.
71. Ibid., 28. See H. M. Dale, 'The Biological Nature of Viruses', *Lancet*, 1931, ii: 601.

72. A. C. Marie, 'Methods of Vaccinating Human Beings Who Have Bitten', League of Nations, *International Rabies*, 36–57.
73. *Report of the Committee of Inquiry on Rabies: Final Report*, Parl. Papers, 1971 [Cmnd. 4696], v, 1021.
74. K. Hummeler and H. Koprowski, 'Investigating the Rabies Virus', *Nature*, 1969, 221: 418–21.
75. S. K. Vernon, A. R. Neurath and B. A. Rubin, 'Electron Microscopic Studies on the Structure of Rabies Virus', *Journal of Ultrastructure Research*, 1972, 41: 29–42.
76. G. M. Baer, *The Natural History of Rabies*, 2 vols, New York: Academic Press, 1975.
77. M. A. Kaplan and H. Koprowski, 'Rabies', *Scientific American*, 1980, 242: 104–113. Z. F. Fu, 'Rabies and Rabies Research: Past, Present and Future', *Vaccine*, 1997, 15: S20–4.
78. M. Kaplan, 'Epidemiology of Rabies', *Nature*, 1969, 221: 421–22.
79. G. S. Turner, 'Extended Quarantine for Rabies', *Nature*, 1969, 224: 246–47.
80. The most detailed account of the incident is a report prepared for Surrey County Council by P. A. Cocking, the Divisional Medical Officer. SZ/CIR/4. CIR/WE/DOC1.
81. *Report of the Committee of Inquiry on Rabies*, 1013. Mrs Marian Hemsley, who had been vaccinated against rabies in India, was awarded the Queen's Medal for Bravery in 1970. *Times*, 13 March 1970, 1d.
82. It was front page news that the Ministry was planning the 'slaughter of the whole animal population as a precaution' using 'Gas Warfare'. *Times*, 25 October 1969, 1c; 28 October 1969, 3a; 1 November 1969, 1b.
83. SZ/CRI/3 CRI5. Report of an enquiry into the origins of the two cases of rabies infection in October and November, MAFF, 1969.
84. Report, 10–11.
85. SZ/CIR/3 CIR 22 'Confidential: Committee of Inquiry on Rabies: Contravention of Quarantine Regulations at Caesar's Camp Kennels', 14 May 1970. The second report could also find no evidence of cross infection at the kennels, but noted that this was unsurprising as there was so much conflicting evidence from the different parties.
86. Later Henderson and Beynon revised their conclusions, stating that 'we have insufficient evidence in the management of the kennels to rule out transmission by direct contact as having been a possibility'. *Report of the Committee of Inquiry on Rabies*, 113–4.
87. *Times*, 5 November 1969, 1d; 17 November 1969, 1b; 5 December 1969, 2d.
88. *Times*, 27 October 1969, 9e.
89. C. K. Pethese and R. Hopc, 'A Case of Rabies at Newmarket', *Veterinary Record*, 1970, 86: 266; *Daily Mirror*, 2 March 1970, 1a–b.
90. *Times*, 2 March 1970, 1g.
91. *Times*, 7 March 1970, 1e, 6g, and 7a.
92. *Daily Mirror*, 7 March 1970, 1d–e.
93. B. Donovan, *Zuckerman: Scientist Extraordinary*, Bristol: BioScientifica, 2005.
94. S. Zuckerman, *Monkeys, Men and Missiles*, London: Collins, 1988.
95. SZ/CIR/ WE/MISC 9; *Sunday Times*, 30 August 1970.
96. *Report of the Committee of Inquiry on Rabies*, Appendix 9, 123–29.
97. *Daily Telegraph*, 25 August 1970, 13; *Times*, 18 August 1979, 1c and 7a; 8 September 1970, 1e.

98. Doomwatch was a science fiction series produced by the BBC. It was broadcast between 1970 and 1972. It centred on a government scientific agency – Department of Observation and Measurement of Scientific Work – that investigated environmental crises.
99. SZ/CIR/16.
100. R. H. Glyn, 'Rabies, Quarantine and Vaccination', *Kennel Club Gazette*, October 1970, 7–8.
101. SZ/CIR/4 CIR/WE/DOC28. H. G. Lloyd, 'Some Observations on Foxes and Rabies'. Pest Infestation Control Laboratory. Mimeo. 30 September 1970.
102. *Daily Telegraph*, 30 June 1971, 2a; *Daily Mail*, 29 June 1971. *Report of the Committee of Inquiry on Rabies*, 10.
103. A. D. Irvin, 'The Epidemiology of Wildlife Rabies', *Veterinary Record*, 1970, 87: 338–48.
104. SZ/CIR/16 Press-cuttings, 1970–1971.
105. *Veterinary Record*, 1973, 93: 60, 273–74, and 489.
106. *Times*, 11 July 1973, 7d and 16 July 1973 13g. Also see *Veterinary Record*, 1973, 93: 549 and 1975, 96: 74.
107. *Times*, 18 September 1974, 3a.
108. There were around 100 illegal landings in Kent in 1973. *Times*, 18 September 1974, 3a.
109. *Times*, 18 May 1976, 5a and 1 June 1976, 3a.
110. *Times*, 16 January 1975, 6a. There were 206 smugglers caught in 1974.
111. *Hansard*, HC, 1974, 870: 1040 and 871: 821–27.
112. *Times*, 23 November 1973, 10f.
113. *Times*, 25 September 1974, 17g and 22 June 1976, 15f.
114. *Times*, 18 September 1974, 3a and 11 July 1975, 4c.
115. *Times*, 5 May 1975, 9e.
116. *Hansard*, HC, 1974–1975, 891: 331–32; 892: 70–71. *Times*, 6 May 1975, 4a and 7 May 1975, 14d.
117. *Times*, 8 June 1975, 1c and 2e; 10 July 1975, 4a.
118. *Times*, 11 July 1975, 4c; *Daily Mirror*, 24 September 1975, 4d–g.
119. National Archives, PREM 16/929, 14–27 January 1976.
120. Wilson left office on 6 April 1976 to everyone's surprise. We now might speculate that his enquiry about Paddy may have been an early sign that he was planning to spend more time with his family and his dog.
121. *Times*, 26 June 1976, 16a.
122. National Archives, MH148 1020, 4A, 12A 16A. *Hansard*, HC, 1975–1976, 906: 468, 477 and 495.
123. At the time of writing, it is possible to view this film on the National Archives website. See http://www.nationalarchives.gov.uk/films/1964to1979/filmpage_outbreak.htm
124. MH148, 1020, 11A, 18 July 1976.
125. *Daily Mirror*, 13 July 1976, 1b–g. *Times*, 9 May 1976, 4a; 13 July 1976, 4a, and 13 August 1976, 3e.
126. *Daily Mirror*, 22 April 1976, 7c–g.
127. *Daily Mirror*, 28 June 1976, 11, d.
128. *Hansard*, HC, 1975–1976, 910: 19–20, 191–92, 351–52 and 488–921; 911: 667–68 and 1707–10.

129. *Times*, 2 May 1976, 1a; 5 June 1976, 4a; 24 August 1976, 3a. *Daily Mirror*, 5 June 1976, 11b.
130. K. Mellanby, 'Shooting Rabbits', *Nature*, 1976, 261: 187.
131. *Times*, 25 June 1976, 16a.
132. *Hansard*, HC, 1975–1976, 911: 381; 913: 503–14.
133. *Daily Mirror*, 28 June 1976, 11d–e.
134. National Archives, 1976, INF 6/2298 and 1983, INF 6/2354.
135. *Times*, 7 June 1976, 3a.
136. Office of Health Economics, *Rabies*, London, 1976.
137. *Times*, 18 June 1976, 16e.
138. *Times*, 22 June 1976, 15f. The government used 'D Notices' to keep topics, usually, matters of state security, out of the public domain.
139. *Times*, 16 June 1976, 15a; 22 June 1976, 15f, 28 June 1976, 14g.
140. *Times*, 28 June 1976, 14g.
141. *Times*, 5 July 1976, 13d.
142. *Times*, 12 July 1976, 13g. However, in the same month the Ministry stocked up with 250,000 doses of vaccine in case of an outbreak.
143. *Times* 29 February 1977, 4f.
144. *Times*, 5 September 1978, 3b.
145. G. N. Henderson and K. White, *Rabies: The Facts You Need to Know*, London: Barrie and Jenkins, 1978.
146. Ibid., p. 128.
147. Also see C. Kaplan et al., eds, *Rabies: The Facts*, Oxford: Oxford University Press, 1977.
148. *Times*, 12 August 1978, 3d.
149. *Times*, 12 August 1978, 3d.
150. Hansard, HL, 1979–80, 986: 211–12. *Times*, 18 June 1980, 4d.
151. This colony of badgers inhabited the Maginot Line – the concrete fortifications constructed by France along its borders with Germany and Italy.
152. Implicit in the narrative is the much lauded British stereotype of Europeans being at ease with adultery.
153. In the same year, British audiences also went to see the cult Canadian director David Cronenberg's second commercial film *Rabid*. Though the film was set in Canada and primarily acted as a critique of the government policy during a city-wide outbreak, it was largely regarded as a critique of martial law. Also, the host of the infection was a woman, who had been left with a lust for human blood, after undergoing experimental skin grafts after a motorcycle accident. When she fed on human victims, who were all usually males and were lured by her 'innocent' looks, she left them with a rabies-like virus that took immediate effect. The fact that the actress was a well known female pornography star undoubtedly deepened the explicit sexual overtones of *Rabid*. Additionally, the film reveals disturbing images of human rabies, the population is rapidly turned into flesh eating menacing zombies – a well developed film genre by this time. Human victims develop hysterical and maniacal symptoms, biting other people or convulsing with laughter, always filled with malicious intent.
154. *Veterinary Record*, 1980, 106: 549. MAFF sponsored an exhibition on rabies at the Science Museum. *Veterinary Record*, 1980, 106: 387.
155. WHO, 'Rabies', http://www.who.int/mediacentre/factsheets/fs099/en [accessed 15 June 2007].

156. *Veterinary Record*, 1980, 106: 139 and 517.
157. See the Parliamentary questions of the leading Eurosceptic Conservative MP Teddy Taylor. *Hansard*, HC, 1985–1986, 87: 268; 90: 401; 92: 198; 98: 366; 102: 808–9.
158. *Veterinary Record*, 1980, 106: 549; *Times*, 16 March 1983, 5f.
159. *Veterinary Record*, 1982, 110: 92; *Times*, 16 March 1982, 5f and 26 May 1982, 5f
160. *Times*, 26 May 1982, 5f; *Veterinary Record*, 1982, 111: 4.
161. P-P. Pastoret et al., eds, *Vaccination to Control Rabies in Foxes*, Brussels-Luxembourg: Office for Official Publications of the European Communities, 1988.
162. *Veterinary Record*, 1986, 119: 413–14.
163. *Times*, 2 December 1985, 1f.
164. Ibid., 6 December 1985, 4a.
165. Ibid., 29 November 1985, 16f.
166. Ibid., 30 November 9a–c.
167. Ibid., 2 December 1985, 1f.
168. N. Tebbit, *The Field*, May 1990.
169. *Veterinary Record*, 1973, 92: 405 and 93: 549.
170. *Veterinary Record*, 1986, 119: 487.
171. Ibid., 118: 223 and 1990, 26: 559.
172. *Daily Mirror*, October 1990, 19.
173. Quoted in E. Darian-Smith, *Bridging Divides: The Channel Tunnel and English Legal Identity in the New Europe*, Berkeley: University of California Press, 1999. J. Barnes, *Letters from London*, New York: Vintage, 1995. Also see *Times*, 6 May 1994.
174. Darian-Smith, *Bridging Divides*, 158–59.
175. Ibid., 147.
176. *Veterinary Record*, 1996, 139: 32–35.
177. D. MacKenzie, 'Europe Launches Spring Offensive Against Rabies', *New Scientist*, 30 April 1994, 6.
178. *Times*, 5 February 1992, 3 and 8 February 1992, 5.
179. *Veterinary Record*, 1992, 130: 21 and 129.
180. http://catalogue.bbc.co.uk/catalogue/infax/programme/RSED702D
181. *New Scientist*, 20 October 1990, 20–21 and 7 March 1992, 9.
182. Ibid., 3 June 1995, 12.
183. *Agriculture Select Committee: Health Controls on the Importation of Live Animals. Fifth Report in 2 Volumes with Proceedings Evidence and Appendices.* Parl Papers, 1993/94, HC 347, Vols. I and II.
184. Ibid., Vol. I, xxiv–xxxii.
185. Ibid., Vol. II, 171–92.
186. Ibid., 189.
187. Ibid., 196.
188. *BMJ*, 1995, 311: 266; *Veterinary Record*, 1995, 137: 130–31.
189. 'Vets in Support of a Change', http://www.dip.deon.co.uk/vet1.htm [Accessed 8 January 2007].
190. *Times*, 20 January 1996, 1; *Evening Standard*, 22 June 1998, 5.
191. *The Independent*, 20 October 1996, 13.
192. *Times*, 24 February 1997, 2. The final ministerial answer on rabies and quarantines by the Conservative government was on 18 March 1997 and was

about bolstering defences at the Channel Tunnel. *Hansard*, HC, 290, 1997–1998, 563–64.

193. *Daily Telegraph*, 8 March 2000.
194. R. Hill, 'Quarantine Reform Campaign Report Published', *Veterinary Practice*, April 1998, 2.
195. *Daily Telegraph*, 24 September 1998.
196. http://www.lowefusion.com/sub_page.cfm/section/casestudies/editID/27 [Accessed 16 January 2007].
197. D. MacKenzie, 'Will Rabies Bite Back?', *New Scientist*, 8 November 1997, 24.
198. *New Scientist*, 5 June 1995, 12.
199. There were complaints about the delay in publication and possible change of mind by the government. *Daily Telegraph*, 30 August 1998 and 20 September 1998.
200. Incidentally, Liz Hurley had starred with Joss Ackland in a movie entitled 'Mad Dogs and Englishmen' in 1995. The film was also released under the title 'Shameless'.
201. Channel 4, 'Undercover Britain – Inside Quarantine', Small World Productions, Broadcast on 5 May 1998. The programme was the subject of a complaint that the filming took place five months before the broadcast and did not reflect the state of the kennels in May 1998. The Independent Television Commission found that the programme makers were in breach of its Programme Code.
202. *Quarantine and Rabies: A Reappraisal*, [PB 3968], London: MAFF Publications, 1998.
203. *New Scientist*, 8 October 1998, 13.
204. *Daily Telegraph*, 24 September 1998.
205. *BBC News*, 6 August 1998.
206. *Daily Telegraph*, 29 February 2000, 1.
207. 'Blair Let's Boy's Best Friend Skip Quarantine', Ibid., 24 July 1999.
208. P. P. Pastoret and B. Brochier, 'Epidemiology and Control of Fox Rabies in Europe', *Vaccine*, 1999, 17, 1750–1754.

Index